MOBILE PHONES AND DRIVING

MOBILE PHONES AND DRIVING

DANIEL M. STURNQUIST
EDITOR

Novinka Books
New York

For permission to use material from this book please contact us:
Telephone 631-231-7269; Fax 631-231-8175
Web Site: http://www.novapublishers.com

NOTICE TO THE READER

The Publisher has taken reasonable care in the preparation of this book, but makes no expressed or implied warranty of any kind and assumes no responsibility for any errors or omissions. No liability is assumed for incidental or consequential damages in connection with or arising out of information contained in this book. The Publisher shall not be liable for any special, consequential, or exemplary damages resulting, in whole or in part, from the readers' use of, or reliance upon, this material.

This publication is designed to provide accurate and authoritative information with regard to the subject matter covered herein. It is sold with the clear understanding that the Publisher is not engaged in rendering legal or any other professional services. If legal or any other expert assistance is required, the services of a competent person should be sought. FROM A DECLARATION OF PARTICIPANTS JOINTLY ADOPTED BY A COMMITTEE OF THE AMERICAN BAR ASSOCIATION AND A COMMITTEE OF PUBLISHERS.

LIBRARY OF CONGRESS CATALOGING-IN-PUBLICATION DATA
Mobile phones and driving / Daniel M. Sturnquist, editor.
    p. cm.
Includes bibliographical references and index.
ISBN 1-60021-162-3
1. Motor vehicles--Law and legislation--United States. 2. Cellular telephone systems--Law and legislation--United States. 3. Traffic safety--United States. I. Sturnquist, Daniel M.
KF2209.M63   2006
343.7309'943--dc22   2006010279

Published by Nova Science Publishers, Inc. ✦ *New York*

CONTENTS

PREFACE

In the United States, as well as worldwide, there has been substantial growth in the use of mobile wireless telecommunication services ("mobile telephones"). The use of mobile telephones by the drivers of motor vehicles has been the subject of certain state and local restrictions that were written because of concerns for driver safety. At the present time, eighteen states and the District of Columbia have enacted legislation concerning the use of mobile telephones by drivers of motor vehicles. The existing state laws vary greatly and are summarized in this new book. Fifty-three pieces of legislation have been introduced in 2004 in twenty-six states and the District of Columbia concerning the use of mobile telephones by drivers of motor vehicles. The current status of state legislation is summarized state by-state in this book. The book includes a bibliography of other books regarding this issue, as well.

In the United States, as well as worldwide, there has been substantial growth in the use of mobile wireless telecommunication services ("mobile telephones"). The use of mobile telephones by the drivers of motor vehicles has been the subject of certain state and local restrictions. S. 179 (108th Cong., 1st Sess. (2003)) has been introduced in the 108th Congress to provide some federal oversight of mobile telephone use by drivers of motor vehicles, by requiring the individual states to enact legislation to restrict mobile telephone use by drivers of motor vehicles. Noncomplying states would be subject to the loss of federal highway funds. At the present time, eighteen states and the District of Columbia have enacted legislation concerning the use of mobile telephones by drivers of motor vehicles. The existing state laws vary greatly and are summarized in the report. Some state laws expressly preempt local regulation of mobile telephones. As of the date of this report, fifty-three pieces of legislation have been introduced in 2004 in twenty-six states and the District of Columbia concerning the use of mobile telephones by drivers of motor vehicles. The current status of state legislation is summarized state-by-state in the chapter. Over the years, Congress has repeatedly conditioned the use of federal highway funds to encourage states to enact desired transportation-related legislation. For example, Congress has used this legislative device in dealing with drunk driving. The pending federal legislation making federal highway funding contingent on state restriction on the use of mobile telephones by drivers of motor vehicles would appear to follow the driving while intoxicated legislative models. Chapter 1 examines the pending federal legislation, pending and enacted state legislation, and the possible effect that the federal legislation, if enacted, might have on the existing state and local regulations.

Research has shown that phone conversations while driving impair performance. It is difficult to quantify the risk of this impairment because the reference is usually made to normal driving without using a phone. "Worse than normal driving" does not necessarily mean dangerous. There is a need to benchmark driving performance while using a mobile phone to a clearly dangerous level of performance. Driving with a blood alcohol level over the legal limit is an established danger. The study in chapter 2 was designed to quantify the impairment from hands-free and hand-held phone conversations in relation to the decline in driving performance caused by alcohol impairment. The TRL Driving Simulator was used to provide a realistic driving task in a safe and controlled environment. Twenty healthy experienced drivers were tested in a balanced order on two separate occasions. The drivers were aged 21 to 45 years (mean = 32, SD = 7.8) and were split evenly by gender. Before starting the test drive, participants consumed a drink, which either contained alcohol or a similar looking and tasting placebo drink. The quantity of alcohol was determined from the participant's age and body mass using the adjusted Widmark Formula (the UK legal alcohol limit 80mg / 100ml).

In: Mobile Phones and Driving
Editor: D. M. Sturnquist, pp. 1-28

ISBN: 1-60021-162-3
© 2006 Nova Science Publishers, Inc.

Chapter 1

MOBILE TELEPHONES AND MOTOR VEHICLE OPERATION [*]

Douglas Reid Weimer

Legislative Attorney, American Law Division

SUMMARY

In the United States, as well as worldwide, there has been substantial growth in the use of mobile wireless telecommunication services ("mobile telephones"). The use of mobile telephones by the drivers of motor vehicles has been the subject of certain state and local restrictions. S. 179 (108th Cong., 1st Sess. (2003)) has been introduced in the 108th Congress to provide some federal oversight of mobile telephone use by drivers of motor vehicles, by requiring the individual states to enact legislation to restrict mobile telephone use by drivers of motor vehicles. Noncomplying states would be subject to the loss of federal highway funds.

At the present time, eighteen states and the District of Columbia have enacted legislation concerning the use of mobile telephones by drivers of motor vehicles. The existing state laws vary greatly and are summarized in the report. Some state laws expressly preempt local regulation of mobile telephones. As of the date of this report, fifty-three pieces of legislation have been introduced in 2004 in twenty-six states and the District of Columbia

---

[*] Excerpted from CRS Report RL31588. Updated September 30, 2004.

concerning the use of mobile telephones by drivers of motor vehicles. The current status of state legislation is summarized state-by-state in the article.

Over the years, Congress has repeatedly conditioned the use of federal highway funds to encourage states to enact desired transportation-related legislation. For example, Congress has used this legislative device in dealing with drunk driving. The pending federal legislation making federal highway funding contingent on state restriction on the use of mobile telephones by drivers of motor vehicles would appear to follow the driving while intoxicated legislative models.

This article examines the pending federal legislation, pending and enacted state legislation, and the possible effect that the federal legislation, if enacted, might have on the existing state and local regulations.

BACKGROUND

In the United States, as well as worldwide, there has been substantial growth in the number of subscribers to mobile wireless telecommunication services.[1] It is reported that there are currently over 169,000,000 mobile telephone[2] subscribers in the United States.[3] The number of subscribers is increasing each year, and this growth is expected to continue.[4]

Concurrent with the increase in mobile wireless telecommunication services ("mobile telephones") has been the use of mobile telephones and related accessories by drivers of motor vehicles.[5] Opinion is divided regarding the desirability of mobile telephone use by drivers. Arguments have been made that mobile telephone use by a motor vehicle operator is a safety feature for summoning emergency personnel and/or roadside assistance. Others argue that the operator's mobile telephone use is a driver distraction that could lead to hazardous situations and to possible accidents.[6]

Certain state[7] and local jurisdictions[8] have enacted and implemented restrictions on the use of mobile telephones by the drivers of motor vehicles. Numerous foreign jurisdictions restrict and/or prohibit mobile telephone use and/or other wireless technology use in motor vehicles.[9]

Federal legislation has been introduced in the 108th Congress — "Mobile Telephone Driving Safety Act of 2003" — which would have the effect of placing some restrictions on the use of mobile telephones by drivers of motor vehicles under certain circumstances.[10] This report examines the pending federal legislation concerning mobile telephone use by the drivers of motor vehicles, pending and enacted state legislation, and the possible effect that the federal legislation, if enacted, might have on the existing state and local laws.

SUMMARY OF LEGISLATION IN THE 108TH CONGRESS

In the 108th Congress, Senator Corzine reintroduced his bill from the 107th Congress[11] as the "Mobile Telephone Driving Safety Act of 2003," S. 179,[12] on January 16, 2003. On that same day, the bill was referred to the Committee on Environment and Public Works, which was the last action on the bill.

Like its predecessor bill, S. 179 would direct the Secretary of Transportation to withhold federal highway funds from any state that permitted an individual to use a mobile telephone while operating a motor vehicle. The bill defines "motor vehicle" to mean a vehicle driven or drawn by mechanical power and manufactured primarily for use on public highways, but does not include a vehicle operated only on a rail.

The bill would withhold federal highway funds from the noncomplying states during their period of noncompliance. The Secretary would be required to withhold 5 percent of federal highway funds from noncomplying States for FY2005. In subsequent fiscal years, the Secretary is required 10 percent of federal highway funds from noncomplying states.

A state would meet the federal requirements if it has enacted and is enforcing a law that makes it unlawful to use a hand-held mobile telephone by an individual operating a motor vehicle. However, the state may permit an individual operating a motor vehicle to use a mobile telephone with a device that permits hand-free operation of the telephone if the state determines that such use does not pose a threat to public safety. Another exception is permitted in the case of an emergency or other exceptional circumstances, as determined by the state.

Federal highway funds withheld from a noncomplying state are to remain available until the end of the fourth fiscal year following the fiscal year for which the funds were authorized to be appropriated. If a noncomplying state complies with federal requirements within the withholding period, withheld funds will be apportioned to it and remain available for expenditure during the following three fiscal years. Any apportioned funds that are not obligated at the end of this period shall be allocated among the states that meet the federal requirements.

S. 179 does not address the issue of the use of mobile telephones by public safety personnel, such as police, firefighters, and emergency medical personnel. While the bill provides an exception for "emergency" situation, it is unclear whether regular use of mobile telephones by these public safety personnel would fall within this exception. Similarly, the bill does not provide a specific definition for the term "mobile telephone." Such definition may be left to the states, but it is possible that such a definition might be interpreted to include other wireless technology. Also, there is no specific definition for the concept of a "hands-free" device. Another issue may be that a headset type feature that covers both ears may prevent the driver from hearing the approach of emergency vehicles and other traffic-related sounds. Some states have laws which prohibit the use of two-earphone devices, whether in use for mobile telephones or for other audio uses. The bill does not address the issue of mobile telephone use by other passengers in the motor vehicle, other than the driver.

SUMMARY OF EXISTING STATE LAWS CONCERNING THE USE OF MOBILE TELEPHONES

At the present time, eighteen states and the District of Columbia have enacted legislation concerning the use of mobile telephones by drivers of motor vehicles. The scope and nature of these laws vary greatly. In 2001, New York became the first state to prohibit drivers from talking on hand-held mobile telephones while operating a motor vehicle.[13] California law requires that rental cars with mobile telephone equipment must include written operating instructions for the safe use of the mobile phone. Another recently enacted California law prohibits the operation of a school or a transit bus while using a mobile telephone. Florida and Illinois permit mobile telephone use, as long as the device does not impair sound to both ears of the driver. Arizona and Massachusetts prohibit school bus drivers from using cell phones while operating a school bus. Massachusetts law requires that all drivers have at least one hand on the steering wheel at all times while using a mobile telephone.

The various state laws concerning mobile telephone use by drivers of motor vehicles are summarized in the following table:[14]

Table 1. State Laws Concerning the Use of Mobile Telephones

State	Rule or Statute	Summary	Penalties
Arizona	Ariz.Admin, Code, Title 17 Chap. 6, Art. 1 R17-9-104 (2002).	A school bus driver may not wear an audio headset or earphones or use a mobile telephone when the school bus is in motion.	No penalty specified.
Arkansas	Ark. Code Ann. § 6-19-120 (2003).	Prohibits the driver use of a mobile telephone while operating a school bus.	Unclassified misdemeanor. Fine of $100 - $250.
California	Cal. Veh. Code § 28090 (West 2001).	Rental cars with mobile telephone equipment must include written operating instructions concerning its safe use.	$100 maximum for first violation; $200 maximum for second; $250 for third and subsequent violations committed within one year.
	A.B. (Assembly bill) 2785 (CA 2004), signed by the Governor on September 14, 2004. To be codified as Chapter No. 505.	Makes it an infraction to drive a school bus or transit vehicle while using a wireless telephone. Contains exceptions.	Driving infraction.
Delaware	H.Con.Res. 30 (2002)(apparently not codified).	Established a task force to study and make findings and recommendations regarding driver distractions, including mobile telephone use.	Not applicable.
	H.B. 379 (DE 2004); Chap. No. 274 (June 24, 2004).	Clarifies that school bus drivers can use a radio or electronic device to make or receive calls for assistance.	Not applicable (amendment to existing law).
	S. B. 174 (DE 2004); Chap. No. 318 (July 6, 2004).	Prohibits the use of a mobile telephone while operating a school bus.	$50 to $100 fine for first offense. $100 to $100 fine and six month bus driving suspension for subsequent offenses.

Table 1. (Continued)

State	Rule or Statute	Summary	Penalties
District of Columbia	D.C. Code §§ 50- 2151 to 2157 (2004).	Mobile telephone use is prohibited by a driver of a motor vehicle, unless equipped with a hands-free accessory. Exceptions for emergency situations and law enforcement and emergency personnel. Total prohibition of mobile telephone use by drivers of a moving school bus carrying passengers; and a person holding a learner's permit.	Fine of $100; provided that the fine shall be suspended for a first time violator who, subsequent to the violation but prior to the imposition of a fine, provides proof of acquisition of a hands-free accessory of the required type/ Violation is to be treated as a moving violation.
Florida	Fla. Stat. Ann. § 316.304 (West 2001).	Mobile telephone use is permitted only if it provides sound through one ear and allows surrounding sound to be heard with the other ear.	$30 for each violation; non-moving violation.
Illinois	625 Ill. Comp. Stat. 5/12-610 (2003).	A single-sided headset or earpiece is permitted with a mobile telephone while driving.	No penalty.
	625 Ill. Comp. Stat. 5/12-813.1 (2003).	Drivers of school buses prohibited from using a mobile telephone while driving; emergency exceptions.	Petty offense; punishable by a $110 to $250 fine.
Kentucky	Ky. Rev. Stat. Ann. § 65.873 (2003).	Prohibits local governments from restricting mobile telephone use by drivers of motor vehicles.	Not applicable.
Louisiana	La. Rev. Stat. Ann. § 33:31 (West 2004).	Local jurisdictions prohibited from regulating mobile telephone use by drivers of motor vehicles.	Not applicable.

Table 1. (Continued)

State	Rule or Statute	Summary	Penalties
Maine	Me. Rev. Stat. Ann. tit. 29-A M.R.S. §§ 1304, 1311 (West 2003).	Persons who have been issued a learners permit may not operate a motor vehicle while operating a mobile telephone.	"Traffic infraction."
		Provides a restricted license for those under age 18; such persons may not operate a motor vehicle while using a mobile telephone.	License restrictions may be extended for periods of time (180 days).
Massachusetts	Mass. Gen. Laws Ann. ch. 90, § 13 (West 2004).	Mobile telephone use is permitted as long as it does not interfere with the operation of the vehicle and one hand remains on the steering wheel at all times.	$35 maximum for first violation; $35 to $75 for second violation; $75 to $150 for third and subsequent violations committed within one year.
	Mass. Gen. Laws Ann. ch. 90, § 7B (West 2004).	No person shall operate a moving school bus while using a mobile telephone. Emergency exception.	No penalty specified.
Mississippi	Miss. Code Ann. § 63-3-212 (2004).	Local jurisdictions are prohibited from enacting ordinances restricting mobile telephone use in motor vehicles.	Not applicable.
Nevada	Nev. Rev. Stat. Ann. § 707.375 (Michie 2004).	Local jurisdictions are prohibited from regulating use of mobile telephones by drivers of motor vehicles.	Not applicable.
New Jersey	N.J. Stat. Ann. § 39:4-97.3 (West 2004).	Use of a hands-free mobile telephone by the operator of a moving motor vehicle is permitted only if "its placement does not interfere with the operation of federally required safety equipment and the operator exercises a high degree of caution in the operation of the motor vehicle."	Not less than $100 and not more than $250.

Table 1. (Continued)

State	Rule or Statute	Summary	Penalties
New Jersey Cont'd	N.J. Stat. Ann. § 39:3B-25 (West 2004)	Drivers of school buses are prohibited from using mobile telephones, except: 1) when the bus is parked in a safe area off highway; or 2) in an emergency situation.	Not less than $250 or more than $500.
New York	N.Y. Veh. & Traf. Code § 1225-c (McKinney 2002).	Drivers are prohibited from talking on hand-held mobile telephones while operating motor vehicles.	Not more than $100.
Oklahoma	H.B. 1081 (2001) (apparently not codified).	Prohibits local jurisdictions from restricting use of mobile telephones by drivers of motor vehicles.	Not applicable.
Oregon	Or. Rev. State. § 801.038 (2003).	A city, county or other local government may not enact or enforce any charter provision, ordinance, resolution or other provision regulating the use of mobile telephones in motor vehicles.	Not applicable.
Rhode Island	R.I. Gen. Laws § 31-22-11.8 (2004).	The use of a mobile telephone by a school bus driver is prohibited while the bus is transporting children, except in the case of an emergency.	$50 fine.
Tennessee	Tenn. Code Ann. § 5508-192 (2004).	School bus driver prohibited from using a hand held mobile telephone while such a vehicle is in motion and is transporting children. Exception for mobile telephone or two-way radio communications made to and from a central dispatch, a school transportation department, or a similar office.	Class C misdemeanor. Fine of $50.

LEGISLATIVE TRENDS

States continue to consider and enact legislation concerning the use of mobile telephones by the drivers of motor vehicles. The NCSL determined that, in 2003 alone, legislators in forty-one states and the District of Columbia considered one hundred sixteen bills concerning mobile telephone use or related distracted driving issues.[15] However, most of these bills were not passed.[16]

To date, no state has outright banned all use of any mobile telephone device while driving. However, there has been a trend toward at least considering a prohibition on the driver use of hand-held telephones. For example, New York, prohibits the use of hand-held telephones while driving except during emergency situations; however, the New York law permits the driver to use hands-free mobile telephone devices. Forty-eight bills which were introduced in thirty-three states considered similar legislation to implement hands-free mobile telephone requirements. None of these bills were enacted into law in 2003, although two such bills have been enacted and implemented in 2004, which are discussed below.

Several states have begun to focus on mobile telephone use by young drivers. Many proposals would restrict use based on age/or type of permit. In 2003, Maine enacted a law to restrict mobile telephone use by drivers under the age of twenty-one who have a learner's permit or a restricted drivers license.

Currently, several states prohibit the drivers of school buses from using mobile telephones while operating a school bus. In 2003, legislatures in six states proposed bills which would implement such restrictions. Arkansas and Tennessee enacted such laws in 2003.

A growing trend is to require the collection of crash data information. Several states currently require law enforcement officers to collect information concerning any connection between mobile telephone use and a motor vehicle crash. Other states have enacted legislation which requires the conduct of studies of the effects of mobile telephone use on traffic safety.

Another legislative theme is for the states to assert authority over the jurisdiction of mobile telephone use and to restrict local mobile telephone use laws. Several states — Kentucky, Louisiana, and Nevada among them — have laws that prohibit local restriction or regulation of mobile telephone use by drivers of motor vehicles.

Certain proposals would place special requirements on the use of mobile telephone use by the drivers of motor vehicles. Such requirements may involve the use of special headsets or hands-free equipment. For example,

California in 2001 enacted legislation that required rental cars equipped with mobile telephone equipment must be furnished with written instructions for the safe use of such equipment.

States have considered other distracted driving legislation concerning distractions to the operator of a motor vehicle. In 2003, Louisiana enacted legislation that prohibited driving a motor vehicle equipped with a televisions capable of receiving any pre-recorded visual presentation, unless the device is behind the driver's seat or is not visible by the driver.

The state legislatures of six states considered measures that would have increased driver responsibility for the involvement in a crash will the driver was using a mobile telephone. None of these bills was enacted.

STATE LEGISLATIVE ACTIVITY IN 2004

In 2004, three jurisdictions enacted legislation concerning the use of mobile telephones by the drivers of motor vehicles: the District of Columbia, New Jersey, and Delaware. The District and New Jersey laws are summarized.

District of Columbia

On January 4, 2004, the District of Columbia City Council approved hands-held restrictions on mobile telephone use by motor vehicle operators as part of the Distracted Driving Safety Act of 2004 ("Act"),[17] an act that also set other standards for "distracted driving."

Beginning on July 1, 2004, it became illegal for drivers to use a mobile telephone or other electronic device while driving in the District of Columbia, unless the telephone or device is equipped with a hands-free accessory. The act is designed to improve traffic safety in the District of Columbia by reducing the number of crashes caused by inattentive drivers who become distracted by the use of phones or other electronic devices.[18] The District police department began issuing warnings on July 1, 2004, and actual ticketing began on August 1, 2004.

The statute specifically states that no one "shall use a mobile telephone or other electronic device while operating a moving vehicle in the District of Columbia, unless the telephone or device is equipped with a hands-free accessory."[19] Exceptions to the prohibition is made for: 1) emergency use of a mobile telephone, including calls to 911 or 311, a hospital, an ambulance service provider, a fire department, a law enforcement agency, or a first-aid squad; 2) use by law enforcement and emergency personnel or by a driver of an authorized emergency vehicle, acting within the scope of official duties; or initiating or terminating a telephone call, or turning the telephone on or off.[20] The driver of a school bus is specifically prohibited from using a mobile telephone, except in cases or emergency or under certain very specific cases.[21] A person holding a learner's permit is prohibited from using any mobile telephone or other device, including those with hands-free accessories, while operating a moving vehicle on a public highway, except in an emergency.[22] No age restriction or limitation specified within the context of the learner's permit.

The statute requires that following a motor vehicle accident, the police officer's report must include whether a mobile telephone was present in a vehicle; whether the use of such telephone contributed to the cause of the accident; and whether any other distraction contributed to the cause of the accident.[23] The Director of the District Department of Transportation is required to annually publish and submit to the City Council a report containing statistics concerning the possible relationship between motor vehicle accidents in the District of Columbia and the use of mobile telephones or other electronic devices by motor vehicle operators.[24] The Mayor is also required to submit a report to the City Council containing recommendations concerning mobile telephone use in motor vehicles.[25]

The statute provides penalty provisions. A fine of $100 is to be imposed for violation of the statutory requirements. However, the fine shall be suspended for a first time violator who, subsequent to the violation but prior to the imposition of the fine, provides proof of a hands-free accessory of the type required by the law.[26]

New Jersey

The New Jersey statute was enacted into law on January 20, 2004, and is more limited in scope than the District of Columbia law. The statute provides that the use of a wireless telephone by the operator of a moving motor vehicle on a public road or highway shall be unlawful, except when the telephone is a hands-free wireless telephone, provided that its placement does not interfere with the operation of federally required safety equipment and the operator exercises a high degree of caution in the operation of the motor vehicle.[27] An exception is allowed to use a hand-held telephone if: the driver has reason to fear for his life or safety, or believes that a criminal act may be perpetrated against himself or another person; or the operator is using the telephone to report a fire, traffic accident, hazard, or other dangerous condition. The penalty for the violation of the New Jersey statute is a fine of not less than $100, and not more than $250.[28]

Other States

In 2004, various states have continued their efforts to regulate the use of mobile telephones by the drivers of motor vehicles. Delaware[29] enacted two laws relating to the use of radios or electronic devices by the drivers of school buses[30] and prohibiting the use of mobile telephones while operating a school bus.[31]

Pending state legislation is summarized in table 2.

LEGAL CONSIDERATIONS RELATED TO THE ENACTMENT OF FEDERAL LEGISLATION

Over the years, Congress has repeatedly conditioned the use of federal highway funds to encourage states to enact desired transportation-related legislation. For example, Congress has used this legislative device in dealing with various issues related to the issue of driving while intoxicated. The Highway Safety Amendments of 1984 effectively established the national minimum drinking age by providing that Congress would withhold 5 percent (increasing to 10 percent) of federal highway funds for a state's failure to enact a minimum drinking age of twenty-one.[32] Similarly, other federal legislation has been enacted that conditions the receipt of federal highway funds on state adoption of federal standards governing the revocation or suspension of drivers' licenses of individuals convicted of drug offenses;[33] the operation of motor vehicles by intoxicated minors;[34] and the establishment of minimum penalties for repeat offenders for driving while intoxicated or driving under the influence.[35] The pending federal legislation making federal highway funding contingent on state restriction on the use of mobile telephones by drivers of motor vehicles would appear to follow these legislative models.

Table 2. 2004 State Legislative Activity Concerning Mobile Telephone Use and Motor Vehicle Operation

State	Bill Number	Bill Summary and Penalties	Status of State Activity
Alabama	H.B. 117 (AL 2004).	Prohibits the use of hand-held mobile telephones while driving. Headsets can only allow sound to one ear. Fines of up to $50.	Referred to the House Public Safety Committee (February 3, 2004).
Arizona	H.B. 2691 (AZ 2004).	Prohibits the use of hand-held telephones while driving. Exceptions provided for emergency situations. Violations have fines of $50. If driver involved in an accident, fine may be $200. Law enforcement required to collect information about mobile telephone use on crash report forms.	Referred to the House Rules Committee (February 16, 2004).
California	A.B. (Assembly bill) 1828 (CA 2004).	Prohibits the use of a hand-held mobile telephone while operating a motor vehicle. Exceptions for emergency situations. Convictions for a first offense are punishable by a $20 fine. Subsequent offenses are punishable for a $50 fine.	Referred to the Assembly Transportation Committee. Not heard (April 19, 2004).
	A.B. (Assembly bill) 2785 (CA 2004). Enacted.	Prohibits operating a school bus or a transit bus while using a mobile telephone.	Signed by the Governor on September 14, 2004. To be codified at Chapter No. 505.
	S.B. 1320 (CA 2004).	Prohibits the use of a mobile telephone while driving in a school zone.	Referred to the Senate Transportation Committee (April 20, 2004).

Table 2. Continued

State	Bill Number	Bill Summary and Penalties	Status of State Activity
Colorado	H.B. 1173 (CO 2004).	Makes it a secondary traffic offense for a holder of a temporary instruction permit or a minor's instruction permit to drive while using a mobile telephone or other mobile communication device (other than a hands-free device). Provides exemptions and penalty assessments.	Passed House and referred to the Senate Committee on Transportation. Postponed indefinitely (March 3, 2004).
Connecticut	H.B. 5553 (CT 2004).	Prohibits the use of hand-held telephones while driving. Prohibits drivers from using a mobile electronic device to perform any personal computer function, sending or receiving any electronic mail, playing any video game or digital video disk, or taking or transmitting any digital photograph while operating a motor vehicle. Fines range from $75 for a first violation, to $150 for a second violation, to $250 for a third and subsequent violation.	Referred to the Joint Judiciary Committee. Failed joint favorable deadline (March 22, 2004).

Table 2. Continued

State	Bill Number	Bill Summary and Penalties	Status of State Activity
Delaware	H.B. 224 (DE 2004).	Prohibits the use of hand-held mobile telephones while operating a motor vehicle and makes it a secondary offense. Contains emergency exceptions. Requires schools to place inattentive driving on the school curriculum. Prohibits use of a mobile telephone while operating a school bus. Preempts local jurisdictions from enacting restrictions on the use of mobile telephones while driving. Requires sellers of mobile telephones to educate customers about improper use of the telephone while driving. Requires the Dept. of Public Safety to collect crash statistics related to mobile telephone use while driving.	Reported out of the Public Safety Committee without recommendation (June 11, 2004).
	H.B. 379 (DE 2004).	Clarifies that school bus drivers can use a radio or electronic device to make or receive calls for assistance.	Signed by the Governor and became Chapter No. 274 (June 24, 2004).
	S.B. 174 (DE 2004).	Prohibits the use of a mobile telephone while operating a school bus. Penalties provided.	Signed by the Governor and became Chapter No. 318 (July 6, 2004).
	S.B. 244 (DE 2004).	Prohibits drivers with only a learner's permit from using a mobile telephone.	Passed Senate. Referred to House Public Safety Committee. Laid on table (June 1, 2004).

Table 2. Continued

State	Bill Number	Bill Summary and Penalties	Status of State Activity
District of Columbia (See discussion above).	B15-0035 (DC 2004).	Prohibits the use of hand-held telephones while driving. Provides exceptions for emergency situations. Violators may be punished with fines of $100. Requires police to collect information on crash reports about mobile telephone use. Requires the Department of Motor Vehicles to publish crash statistics concerning the relationship between mobile telephone use and motor vehicle crashes.	Enacted on January 6, 2004.
Georgia	H.B. 1241 (GA 2004).	Prohibits restricted drivers (Class D) from using mobile telephones while operating a motor vehicle.	Referred to the House Motor Vehicles Committee (January 29, 2004).
Illinois	H.B. 5020 (IL 2004).	Provides that a person who holds an instruction permit, or a person who has held a driver's license for less than one year, may not use a mobile telephone while driving a vehicle. Provides that a person who is not subject to those prohibitions may use a mobile telephone while driving if he or she obeys all traffic laws. Provides that if a person permitted to use a wireless telephone commits a traffic violation while using a wireless telephone, he or she, in addition to other violations, is guilty of a petty offense punishable by a fine of not more than $79.	Tabled by sponsor (March 25, 2004).

Table 2. Continued

State	Bill Number	Bill Summary and Penalties	Status of State Activity
Illinois Continued		Provides that a violation of the provision or a similar provision of a local ordinance is an offense against laws or ordinances regarding the movement of traffic. Contains emergency exceptions.	
	H.B. 6568 (IL 2004).	Provides that a person under age 19 may not use a mobile or other telephone while driving. Contains emergency exceptions. Violation is a petty offense punishable by a fine of $100.	Referred to the House Rules Committee (February 6, 2004).
	H.B. 6636 (IL 2004).	Makes it a petty offense to drive a motor vehicle while using a mobile telephone, unless that telephone is designed for hands- free operation and is used in that manner while driving. Fine of not more than $20 for a first offense and not more than $50 for each subsequent offense. Contains emergency exceptions.	Referred to the House Rules Committee (February 9, 2004).
	S.B. 2575 (IL 2004).	Provides that a person who holds an instruction permit or a graduated license may not use a mobile telephone while driving. Contains emergency exceptions.	Referred to the House Rules Committee (February 4, 2004).

Table 2. Continued

State	Bill Number	Bill Summary and Penalties	Status of State Activity
Indiana	S.B. 131 (IN 2004).	Makes it a Class B infraction with a fine of $1,000 to operate a motor vehicle and simultaneously use a mobile telephone, except in emergencies. Authorizes a person who views the operation of a vehicle with simultaneous mobile telephone use and driving on certain highways to report the incident to the state police or sheriff. Requires the state police or sheriff to issue a notice to the registered owner of the motor vehicle.	Referred to the Committee on Criminal and Civil and Public Policy (January 6, 2004).
Iowa	H.B. 2158 (IA 2004).	Prohibits the use of a hand-held mobile telephone while driving. There is a $25 fine upon conviction.	Referred to the House Transportation Committee (February 4, 2004).
Kentucky	H.B. 602 (KY 2004).	Prohibits the use of hand-held mobile telephones while operating a motor vehicle. Requires the Department of Transportation to study the effects of the use of mobile telephones and similar equipment and driver distraction on traffic safety. Department must submit a report to the Governor and the legislature in four years. Pre-empts local laws.	Referred to House Transportation Committee (February 26, 2004).
Maryland	H.B. 1151 (MD 2004).	Prohibits minors under the age of 18 from using any mobile telephone while driving. Excepts emergencies. Authorizes a maximum fine of $500 for violations.	Reported unfavorably from the House Committee on Environmental Matters (March 15, 2004).

Table 2. Continued

State	Bill Number	Bill Summary and Penalties	Status of State Activity
Maryland Continued	H.B. 1152 (MD 2004).	Prohibits drivers from engaging in distracting activity including use of a mobile telephone or other electronic device. Excepts hands-free devices. Excepts emergencies. Requires information in accident reports. Requires report to the legislature regarding mobile phone use and auto crashes.	Reported unfavorably from House Environmental Matters Committee (March 22, 2004).
	H.B. 189 (MD 2004).	Prohibits holders of learner's instructional permits or provisional driver's licenses from using specific interactive wireless communication devices while operating a motor vehicle. Excepts emergencies.	Reported unfavorably from the House Committee on Environmental Matters (March 1, 2004).
	H.B. 275 (MD 2004).	Provides that evidence of a motor vehicle driver's operation of a hand-held telephone may be considered by the trier of fact in determining whether the driver was negligent under specified circumstances. Makes specific exceptions.	Referred to House Judiciary Committee. Reported unfavorably (March 15, 2004).
	H.B. 29 (MD 2004).	Prohibits hand-held telephone use while driving. Makes specific exceptions. Requires that a violation be enforced only as a secondary action. Violation of the statute may not be classified as a moving violation for assessing points.	Environmental Matters Committee reported unfavorably (February 13, 2004).

Table 2. Continued

State	Bill Number	Bill Summary and Penalties	Status of State Activity
Maryland Continued	H.B. 5 (MD 2004).	Requires the Motor Vehicle Administration to impose a restriction on learners' instructional permits and provisional drivers' licenses that prohibits permit holders or licensees from using a specified wireless communication device while operating a motor vehicle. Makes specified exceptions.	Reported unfavorably from the House Committee on Environmental Matters (March 15, 2004).
	S.B. 630 (MD 2004).	Prohibits minors from using a mobile telephone while driving.	Referred to the Senate Judicial Proceedings Committee. Reported unfavorably (March 1, 2004).
Michigan	H.B. 5084 (MI 2003).	Prohibits an individual under18 years of age who is driving on a permit or in graduated licensing status from using a hand- held mobile telephone while operating a motor vehicle.	Referred to the Committee on Transportation (September 25, 2003).
	H.B. 5085 (MI 2003).	Prohibits the use of a mobile telephone while driving.	Referred to the Committee on Transportation (September 25, 2003).
Minnesota	H.B. 2712 (MN 2004).	Prohibits operation of a mobile telephone in a moving motor vehicle by the holder of a provisional driver's license or instruction permit.	Referred to the Committee on Transportation Policy (March 3, 2004).
	S.B. 2805 (MN 2004).	Prohibits operation of a mobile telephone in a moving motor vehicle by the holder of a provisional driver's license or instruction permit.	Referred to the Senate Committee on Finance (April 26, 2004).

Table 2. Continued

State	Bill Number	Bill Summary and Penalties	Status of State Activity
Nebraska	L.B. 1111 (NE 2004).	Makes changes regarding holders of provisional operator's permits; prohibits certain forms of mobile telephone use; sets forth penalty provisions.	Referred to Legislative Committee on Transportation and Telecommunications. Hearing notice (February 10, 2004).
New Jersey (See discussion above.)	A.B.(Assembly bill) 3159 (NJ 2004).	Requires driver distraction (including mobile telephone use) to be noted in traffic accident reports.	Introduced and filed (June 24, 2004).
	A.B. (Assembly bill) 664; 965 (NJ 2004).	Prohibits the use of mobile telephones while driving.	Referred to the Assembly Transportation Committee (January 13, 2004).
	S.B. 338 (NJ 2004).	Prohibits the use of a mobile telephone while operating a motor vehicle. Permits hands-free devices. Requires the DMV to collect data on crash report forms.	Enacted January 20, 2004.
New York	A.B. (Assembly bill) 3675 (NY 2004).	Provides that drivers who, while using a mobile telephone, cause accidents which result in serious injury or death shall be subject to the same criminal penalties as drivers who cause serious injury or death while driving under the influence of alcohol or drugs. Imposes a penalty of two points against a person's license when convicted of a violation of driving while using a mobile telephone.	Referred to the Assembly Codes Committee (February 10, 2004).

Table 2. Continued

State	Bill Number	Bill Summary and Penalties	Status of State Activity
New York (cont.)	A.B. (Assembly bill) 4114 (NY 2004).	Requires that police motor vehicle crash reports include information about whether mobile telephones were present in the vehicles and whether the mobile telephones were a contributing factor to the crash.	Referred to the Assembly Transportation Committee (January 7, 2004).
	A.B. (Assembly bill) 5689 (NY 2004).	Prohibits drivers under the age of 18 from using a hands-free mobile telephone while operating a motor vehicle.	Referred to the Assembly Transportation Committee (March 3, 2004).
	A.B. 6379 (Assembly Bill) 6379 (NY 2004).	Prohibits drivers from using hands- free mobile telephones while operating a motor vehicle.	Referred to the Assembly Transportation Committee (March 4, 2004).
	S.B. 3521 (NY 2004).	Amends existing law so that drivers receive one point on their driving records if they are convicted of using a mobile telephone while operating a motor vehicle.	Amended in the Senate Transportation Committee (January 23, 2004).
Rhode Island	H.B.7065 (RI 2004).	Prohibits the use of hand-held mobile telephones while driving a motor vehicle or a bicycle. Fines are $35 for a first offense; $70 for a second offense; and $140 for a third or subsequent offense.	Referred to the House Corporations Committee (January 8, 2004).
	H.B. 7107 (RI 2004).	Prohibits minors, persons under the age of 18, from using a mobile telephone either while operating a motor vehicle or as a passenger. $50 fine.	Referred to the House Judiciary Committee (January 8, 2004).

Table 2. Continued

State	Bill Number	Bill Summary and Penalties	Status of State Activity
South Carolina	H.R. 4412 (SC 2003).	Provides that a person who possesses a beginner's permit may not operate a motor vehicle while using a mobile telephone or other wireless communications device.	Referred to the Committee on Education and Public Works (January 13, 2004).
	H.R. 4703 (SC 2004).	Prohibits the use of hand-held telephones while operating a motor vehicle. Convictions for violations are punishable by a fine of $50 and for imprisonment of not more than 10 days.	Referred to the House Judiciary Committee (February 4, 2004).
South Dakota	S.B. 126 (SD 2004).	Requires the DMV to collect information about mobile telephone involvement in motor vehicle crashes. Requires the Department of Public Safety and the Division of Insurance to evaluate whether the use or nonuse of mobile telephones by motorists should be a factor in insurance premiums, tort liability, and safety instructions.	Tabled in Senate Transportation Committee (January 29, 2004).
	S.B. 136 (SD 2004).	No person with a restricted minor's permit may operate a motor vehicle while using a mobile telephone or other wireless telecommunication device. Emergency exception. The term, for the purpose of the statute, means the talking, listening, or placing or receiving a call on any mobile telephone or other wireless telecommunication device, or operating its keys, buttons, or other controls.	Deferred legislative action (January 29, 2004).

Table 2. Continued

State	Bill Number	Bill Summary and Penalties	Status of State Activity
Tennessee	H.B. 3306 (TN 2004).	Prohibits the drivers of trucks with gross vehicle rating over 16,000 pounds from using a hand-held telephone while driving. $50 fine.	Referred to the House Transportation Committee (February 12, 2004).
	S.B. 2293 (TN 2004).	Prohibits drivers with a learner's permit, an intermediate license, or a restricted license from using a mobile telephone while driving. $50 fine.	Passed Senate (March 30, 2004).
Utah	H.B. 190 (UT 2003).	Prohibits school bus drivers from using a mobile telephone while driving a school bus. Exceptions: medical emergency, safety hazard, or criminal activity.	Amended (February 4, 2004).
Vermont	H.B. 575 (VT 2004).	Prohibits the use of hand-held mobile telephones while driving.	Referred to the House Transportation Committee (January 14, 2004).
	S.B. 199 (VT 2004).	Prohibits the use of hand-held mobile telephones while operating a motor vehicle. Emergency exceptions.	Referred to the Senate Transportation Committee (January 6, 2004).
Virginia	S.B. 581 (VA 2004).	Prohibits the use of hand-held mobile telephones while operating a motor vehicle. $100 fine.	Passed by indefinitely by the Senate Transportation Committee (January 22, 2004).
Wyoming	H.B. 76 (WY 2004).	Prohibits the use of a mobile telephone (cellular or satellite) while operating a motor vehicle.	Introduced (February 9, 2004).

As the Supreme Court has stated: "Congress has frequently employed the Spending Power to further broad policy objectives by conditioning receipt of federal moneys upon compliance by the recipient with federal statutory and administrative directives. The Court has repeatedly upheld against constitutional challenge the use of this legislative device to induce governments and private parties to cooperate voluntarily with federal policy."[36] Various criteria concerning Congress' discretion have been provided by the Court. The conditions placed upon the receipt of funds, like the spending itself, must advance the general welfare, but the decision of that rests largely, if not wholly, with Congress.[37] Since the states may choose to receive or not to receive the proffered federal funds, Congress must set out the conditions unambiguously, so that the states may rationally decide.[38] The Court has suggested that the conditions must be related to the federal interest for which the funds are expended.[39] Furthermore, the power to condition funds may not be used to induce the states to engage in activities that would be unconstitutional.[40]

If the state accepts the federal funds on conditions, and then does not follow the prescribed federal requirements, the typical remedy is federal administrative action to terminate the funding and to recoup funds that the state has already received.[41] It has also been determined that under certain circumstances the recipients and the potential recipients in a particular program may sue to compel states to observe the standards.[42]

CONCLUSION

Along with the substantial growth in mobile telephone use has been concern about the use of mobile telephones by the drivers of motor vehicles. At the present time, a variety of state laws and local ordinances provide some regulation of mobile telephone use. In addition, a number of states are considering legislation to regulate mobile telephone use by drivers of motor vehicles.

Federal legislation has been introduced in the 108th Congress which would effectively require individual states to enact legislation to restrict mobile telephone use by drivers of motor vehicles. Noncomplying states would be subject to the loss of federal highway funds. The legislation appears similar to existing legislative precedent directing states to enact certain legislation or be subject to the loss of federal highway funds.

REFERENCES

[1]   See CRS Report RS20664, Third Generation ("3G") Mobile Wireless Technologies and Services; and CRS Report RS20993, Wireless

Technology and Spectrum Demand: Third Generation (3G) and Beyond.

[2] For the purpose of this report, the term "mobile telephone" is used as a generic term for any type of cellular or wireless telephone.

[3] Figure reported by the Cellular Telecommunications & Internet Association (CTIA). See [http://www.wow-com.com/]. In 2002, the CTIA had reported that there were 137,000,000 subscribers. This indicates an average growth rate of over 20%, each year, in the past two years. The CTIA updates its usage figures regularly.

[4] Id.

[5] Technological advances in wireless telecommunications are combining with Internet technology to develop new generations of applications and services. Currently, the United States and other countries are moving toward a third generation of mobile telephony, known as 3G. The dominant feature of 3G technology is that the transmission speeds are significantly faster. See CRS Report RS20993 at 1. A related development is in the use of wireless fidelity technology, known as Wi-Fi. There are wireless networks which provide high-speed access to the Internet. 3G technology can be described as bringing Internet capability to wireless mobile telephones. Wi-Fi provides wireless Internet access for portable computers and handheld devices, such as Personal Digital Assistants. These technological advances have increased the communications technology available in motor vehicles. In addition to making and receiving calls (other features include call forwarding, paging features, Voicemail, and prioritizing of calls), modern mobile telephones can take, send, and receive pictures. Mobile telephones permit users to surf the World Wide Web, check stock quotes or sports scores, play video games, and perform a variety of other functions, in addition to conversation. Modern motor vehicles may include other technological devices such as televisions, navigation systems, fax machines, dvd players, computers, and other devices. It is expected that additional features will be available for

[6] Mobile telephones, as well as other entertainment/communication devices for use in motor vehicles. This report is limited to the use of mobile telephones by the drivers of motor vehicles.

[7] M. Sundeen, Cell Phones and Highway Safety: 2003 State Legislative Update, (2004) [Report from the National Conference of State Legislatures] at [http://ncsl.org/print/transportation/cellphoneupdate12-03.pdf]. (Cited to afterward as "Sundeen").

[8] State legislative activity is discussed infra.

[9]   The National Conference of State Legislatures ("NCSL") reports that twenty-five municipalities or counties have passed restrictions on the use of mobile telephones while operating a motor vehicle. See Sundeen at 15-16. These jurisdictions, listed alphabetically by state, are: Miami-Dade County, FL; Pembroke Pines, FL; Westin, FL; Brookline, MA; Bloomfield, NJ; Carteret, NJ; Hazlet, NJ; Irvington, NJ; Marlboro, NJ; Nutley, NJ; Paramus, NJ; Santa Fe, NM; Nassau County, NY; Suffolk County, NY; Westchester County, NY; Brooklyn, OH; North Olmstead, OH; Walton Hills, OH; Conshocken, PA; Lower Chichester, PA: West Conshocken, PA; Lebanon, PA; Hilltown Township, PA; York, PA; and Sandy, UT. In 2001, the NCSL had reported thirteen jurisdictions that restricted mobile telephone use by drivers of motor vehicles, which represents nearly a 50% increase in two years. However, it should be noted that although these jurisdictions may have laws/ordinances which limit or restrict mobile telephone use by drivers of motor vehicles, such restrictions may not be strictly enforced, or may not be enforced at all.

[10] The NCSL reports (see Sundeen at 16) that certain restrictions have been placed on mobile telephone use by drivers of motor vehicles in forty-two countries: Australia; Austria: Belgium; Botswana; Brazil; Chile; Czech Republic; Denmark; Egypt; Finland; France; Germany; Greece; Hungary; India; Ireland; Israel; Italy; Japan; Jordan; Kenya; Malaysia; Netherlands; Norway; the Philippines; Poland; Portugal; Romania; Russia; Singapore; Slovak Republic; Slovenia; South Africa; South Korea; Spain; Sweden; Switzerland; Taiwan; Turkey; Turkmenistan; United Kingdom; and Zimbabwe. It is uncertain as to how stringently the restrictions are enforced. It should also be noted that use may be limited by drivers, as well as passengers in motor vehicles.

[11] S. 179, 108th Cong., 1st Sess. (2003).

[12] S. 927, 107th Cong., 1st Sess. (2001).

[13] 108th Cong., 1st Sess.

[14] N.Y. Veh. & Traf. Code § 1225-c (McKinney 2002)..

[15] Table prepared from information derived from Sundeen at 3.

[16] Sundeen at 4-11.

[17] Id.

[18] D.C. Law 15-311 (effective Mar. 30, 2004); codified at D.C. Code §§ 50-2151 to 2157 (2004).

[19] The D.C. Metropolitan District of Columbia Police Department website
      contains   a   comprehensive   summary   of   the   act.   See
      [http://mpdc.gov/info/traffic/distracteddriver.shtm].

[20] Id. § 50-2153(a).

[21] Id. § 50-2153(b).

[22] Id. § 50-2154(a).

[23] Id. § 50-2154(b).

[24] Id. § 50-2156(a).

[25] Id. § 50-2158(a).

[26] Id. § 50-2158(b).

[27] N.J. Stat. Ann. § 39:4-97.3(a) (West 2004).

[28] Id. § 39:4-97.3(d).

[29] See references on chart of state legislative activity in 2004.

[30] H.B. 379 (DE 2004) to be codified at Del. Code Ann. Ch. 274.

[31] S.B. 174 (DE 2004) to be codified at Del. Code Ann. Ch. 318.

[32] 23 U.S.C. § 158.

[33] 23 U.S.C. § 159.

[34] 23 U.S.C. §161.

[35] 23 U.S.C. § 164.

[36] Fullilove v. Klutznick, 448 U.S. 448, 474 (1980)(Chief Justice Burger
      announcing judgment of the Court).

[37] South Dakota v. Dole, 480 U.S. 203, 207 (the placing of conditions on
      the receipt of federal highway funds)(1987).

[38] Id. at 210-211.

[39] Bell v. New Jersey, 461 U.S. 773 (1983); Bennett v. New Jersey, 470
      U.S. 632 (1985); Bennett v. Kentucky Dept. of Education, 470 U.S.
      656 (1985).

[40] King v. Smith, 392 U.S. 309 (1968).

In: Mobile Phones and Driving
Editor: D. M. Sturnquist, pp. 29-64
ISBN: 1-60021-162-3
© 2006 Nova Science Publishers, Inc.

Chapter 2

HOW DANGEROUS IS DRIVING WITH A MOBILE PHONE? BENCHMARKING THE IMPAIRMENT TO ALCOHOL[*]

*Peter Burns, Andrew Parkes, Sue Burton, Rachel Smith and Dominic Burch**

Transport Research Laboratory Crowthorne, Berkshire,
RG45 6AU, UK
*Direct Line Group, Croydon, Surrey, CR9 1AG, UK

SUMMARY

Research has shown that phone conversations while driving impair performance. It is difficult to quantify the risk of this impairment because the reference is usually made to normal driving without using a phone. "Worse than normal driving" does not necessarily mean dangerous. There is a need to benchmark driving performance while using a mobile phone to a clearly dangerous level of performance. Driving with a blood alcohol level over the legal limit is an established danger.

This study was designed to quantify the impairment from hands-free and hand-held phone conversations in relation to the decline in driving

---

[*] Excerpted from www.trl.co.uk

performance caused by alcohol impairment. The TRL Driving Simulator was used to provide a realistic driving task in a safe and controlled environment. Twenty healthy experienced drivers were tested in a balanced order on two separate occasions. The drivers were aged 21 to 45 years (mean = 32, SD = 7.8) and were split evenly by gender. Before starting the test drive, participants consumed a drink, which either contained alcohol or a similar looking and tasting placebo drink. The quantity of alcohol was determined from the participant's age and body mass using the adjusted Widmark Formula (the UK legal alcohol limit 80mg / 100ml).

The test drive had four conditions:

1) motorway with moderate traffic,
2) car following,
3) curving road
4) and dual carriageway with traffic lights.

During each condition the drivers answered a standard set of questions and conversed with the experimenter over a mobile phone. The independent variables in this repeated measures study were normal driving, alcohol impaired driving, and driving while talking on hands-free or hand-held phone.

Results showed a clear trend for significantly poorer driving performance (speed control and response time) when using a hand-held phone in comparison to the other conditions. The best performance was for normal driving without phone conversations. Hands-Free was better than hand-held. Driving performance under the influence of alcohol was significantly worse than normal driving, yet better than driving while using a phone. Drivers also reported that it was easier to drive drunk than to drive while using a phone.

It is concluded that driving behaviour is impaired more during a phone conversation than by having a blood alcohol level at the UK legal limit (80mg/100ml).

INTRODUCTION

The use of mobile telephones in motor vehicles has been associated with a significant increase in the risk of crashing (Redelmeier & Tibshirani, 1997). This has been supported by a body of experimental research that has clearly shown mobile phone use impairs driving performance (see McKnight

& McKnight, 1991; Goodman et al. 1997; and the Stewart Report, 2000 for comprehensive reviews of the research literature). The danger of phone use while driving is that it distracts the driver by taking their attention away from the task of driving. The driver's primary task is to monitor and control the vehicle's lateral and longitudinal position along a safe path. Distracted drivers become dangerous when they are unable to properly monitor and control the vehicle's safe path while using the phone. The typical objective consequence of phone use while driving is poorer lane keeping, more variable speed and a slower reaction time to hazards (e.g., Brookhuis et al., 1991; Fairclough et al., 1991).

Although experimental research has convincingly shown that phone conversations impair driving performance, it is difficult to quantify the risk of this impairment because the reference is usually to normal driving without using a phone. "Worse than normal driving" does not necessarily mean dangerous. There is a need to benchmark driving performance while using a mobile phone to a clearly dangerous level of performance. Society considers driving with a blood alcohol level over the legal limit to be too dangerous. It makes intuitive sense that alcohol impairs driving performance and this was established scientifically nearly 40 years ago (Borkenstein et al.'s Grand Rapids Study, 1964).

Stevens and Paulo (1999), in a TRL Report reviewing the research on phone use in cars, recommended that consideration be given to relating hands-free phone use to alcohol consumption. Although it is an established approach to use drunk driving as a reference to quantify the risks of driving while drowsy (Fairclough & Graham, 1999), no direct comparison has been made to investigate the impairment in driving from phone use. There have been some indirect comparisons to the risks of phone use while driving to drunk driving. Lamble et al (1999) found that engaging in a demanding phone conversation while driving slowed brake reaction times significantly. They found reaction times to be approximately three times that of drivers with a blood alcohol level of 0.05% (less than the UK limit of 0.08%). Redelmeier & Tibshirani (1997) also mention how the relative risk of motor vehicle collision associated to mobile phone use compared to the hazards associated with driving with a blood alcohol level over the legal limit.

The aim of the proposed study is to quantify the distraction from hands-free and hand-held phone conversations in relation to the decline in driving performance caused by alcohol impairment.[It is hypothesised that driving performance decreases more with hand-held phones than hands-free phones. It is also hypothesised that some measures of driving performance while talking on a handheld phone will be significantly worse than driving performance while impaired by alcohol.⌐

METHOD

Subjects

Twenty experienced drivers, aged between 21 and 45 years (mean = 32, SD = 7.8), participated in this study. The sample was split evenly by gender. All were healthy and regular mobile phone users. Participants were randomly selected from the TRL volunteer database, a pool of 1300 drivers representing a cross-section of the driving population. Drivers were paid 70 pounds for their participation in this experiment.

The participants were informed in writing about the study's treatment conditions as well as their right to withdraw at any time. They were reminded that participation was voluntary and based on their informed consent. Upon completing the whole experiment, the participants were debriefed as to the aims of the study, and any questions they had about the study were answered. Ethical approval was obtained from an Ethics Review Committee prior to the start of this study.

Figure 1. Side View of the TRL Driving Simulator

Driving Simulator

The TRL Driving Simulator consists of a medium sized saloon car surrounded by large projection screens giving 210-degree horizontal and 40-degree vertical front vision, and 60-degree horizontal and up to 40-degree vertical rear vision, enabling the normal use of all vehicle mirrors (see Figure 1). The road images are generated by advanced graphic workstations and projected at life size onto the screens. The car body shell incorporates hydraulic rams that supply motion to simulate the heave, pitch and roll experienced in normal braking, accelerating and cornering. Also, when negotiating curves, the simulator provides realistic forces experienced by the driver through the steering wheel. The realism of the driving experience is further enhanced by the provision of car engine noise, external road noise, and traffic sounds.

Route and Traffic Scenarios

Participants drove a 15 km route that was composed of four different segments. The route started with a car following task on a motorway. Drivers were instructed to maintain their present distance from the lead vehicle (54 m), that is the distance between vehicles at the start of the scenario when they were stationary. The car following task lasted for 3.5 km. The lead vehicle oscillated its speed between 50 and 70 mph. The list of instructions is in Appendix A.

After completing the car-following task, drivers were instructed to drive as they would normally on a motorway. The motorway had 3 lanes and a speed limit of 70 mph (113 km/h), the standard speed for motorways in the UK. There was a moderate amount of traffic on the motorway. The traffic was programmed to vary their speeds in relation to the subject's vehicle and could overtake or be overtaken depending on how the subject drove. The motorway continued for 4.7 km.

A section of curved road was used to measure the driver's ability to control the vehicle on a more demanding type of rural road. The curves forming two loops and full length of 3.6 km including the straight segments connecting the loops. The loops were modelled after the TRL research track and each had a changing radius. Drivers were instructed to maintain a speed of 60 mph (96.6 km/h).

This was followed by a 3.3 km section of dual carriageway (2 lane road), which ended in a traffic light. During this section drivers had to respond to 12 warning signs at various points along the dual carriageway. They were instructed to flash their headlights whenever a target sign appeared. There were 4 different warning signs in this choice reaction time task: Elderly pedestrians, Pedestrian crossing, Cyclists and Roadworks. The warning signs were spaced approximately 225 metres apart on average.

Procedure

The participants were invited to the laboratory for a pre-trial session. During this session they were asked to provide background information on their driving history and health. They were given a brief description of the experiment, their height and weight was measured and they were asked to sign a consent form. This description did not mention the specific aims of the experiment, only the treatment conditions. Participants were introduced to the simulator and given a test drive to allow them to become more familiar and comfortable with the environment.

The next trial was scheduled within a week of their familiarisation drive, and the second and final trial was run one week later. Both test trials started off with a familiarisation drive in the simulator where they had a chance to practice the driving tasks.

During the test trials, participants were asked to drive as they normally would and to converse with the experimenter whenever prompted to do so. The routes were driven three times for each visit to the simulator. During the phone visit, drivers were asked to drive the route as they would normally (Control), drive while talking on a handheld phone (Hand held), and drive while talking on a hands-free phone (Hands free). The order of these conditions was randomised. Before driving, participants also had a similar conversation with the experimenter while seated at a table. This was used as a control condition to evaluate conversation performance while driving. On the alcohol visit, they drove the route three times without using the phone.

All participants were breath-tested when they arrived for both sessions and before undertaking the each of the test drives. Before starting with the test drive, participants consumed a drink, which eithercontained a measured amount of alcohol with a disguising mixer or a similar looking and tasting placebo drink. They were given 10 minutes to consume their drink, during which time they were supervised by a researcher. The drink consisted of cream soda either with or without the alcohol (Vodka at 40% alcohol).

The quantity of alcohol within the treatments was determined from the participant's age and body mass using the adjusted Widmark Formula, to achieve the dosing level (Watson et al, 1980). The participant's height and weight was measured to ensure that the participants body mass index (BMI) is within the normal range, and in order to calculate the amount of alcohol required for their dose levels. The drivers were moderately impaired at the UK legal alcohol limit (80mg / 100ml). The average dosage of alcohol for the male drivers was 118 ml, approximately 5 units of alcohol. The average dosage of alcohol was 82 ml for the female drivers, approximately 3 units of alcohol. Participants were breathalysed before each of the three drives to confirm their breath alcohol levels were at or over the limit.

Although the quantity of alcohol dosage varied, the volume of the carbonated mix was adjusted to maintain a constant strength of 20% alcohol by volume. Following the 10 minutes of drinking, the participants waited a further 20 minutes before starting their three test drives. On average, it took approximately 45 minutes to complete the driving. This meant they were driving from between 20 to 65 minutes after dosing. The effects of alcohol may appear within 10 minutes after consumption and peak at approximately 40 to 60 minutes.

Video and audio recordings were taken of the conversations and tasks. The computer running the simulation automatically captured and logged all of the driving behaviour. Subjective workload measures were also taken using the rating scale for mental effort at the end of each of the conditions. Upon completing the whole experiment, the participants completed a questionnaire about their phone usage and attitudes towards phone use while driving (see case report form in Appendix C).

Phone Conversations

It was difficult to design a conversation task that has a consistent level of difficulty within and between drivers. The conversation materials that have been used in car phone research often involve intelligence test type materials (e.g., mathematical computations; Brown, Tickner & Simmonds, 1969) that may represent both extreme and different cognitive loads in relation to normal cellular telephone communications (Goodman et al., 1997). The relevance of this research to normal cellular telephone communications is unclear. Goodman and his colleagues also point out that these studies have used conversations that were free of emotional content (e.g., an argument with a spouse). Discourse that involves substantial degrees of personal involvement may be even more disruptive to driving than the cognitively challenging materials typically included in mobile phone research.

Goodman et al., (1997) identified the need for research to better understand naturalistic driver behaviour while using a cellular telephone. Information on the circumstances of call initiation, call frequency, call length, and call content would help to formulate more realistic test protocols for cellular telephone research. Given that this information is not available, it is difficult to replicate the 'typical' passenger or carphone conversations. Several options are available. One option would be to have drivers speak a person they would normally call (e.g., a friend), however this more natural approach would introduce excessive variability. The same problem of variability could be said about using a negotiation task over the phone (e.g., Parkes, 1991).

Another option would be to have loosely defined conversations of varying levels of complexity. For example: demanding spatial memory question: "how do you get to Heathrow from here", or "describe the waiting room in detail"? Moderate: "what has been in the news lately"? Light: "described your last vacation", or "describe your house"? Although these conversations might be more naturalistic, some work would need to be done to quantify the distraction and make them consistent within and between the drivers.

It was decided to use a script to facilitate the conversation between the subject and driver. This included questions from the Rosenbaum Verbal Cognitive Test Battery (RVCB: Waugh et al., 2000). The RVCB measures judgement, flexible thinking and response times. This has a 30 item remembering sentences task (e.g., repeat the sentence: "Undetected by the sleeping dog, the thief broke into Jane's apartment) and 30 verbal puzzle tasks (e.g., Answer the question: "Felix is darker than Antoine. Who is the lighter of the two"). The test battery has five levels of difficulty with six items within each level of both tests. These questions were split across the conditions and included with lighter and more casual discussions (e.g., a 40 second monologue about a recent holiday). See Appendix B for a list of the conversation questions.

This approach was considered best for the present study because the script was more like a conversation than the mathematical computations used in previous work. Also, with its many questions rated at the same level of difficulty, comparisons could be readily made across the different mediums (hands-free/ hand-held) and between drivers. As an additional check, participants were asked to rate the subjective mental effort required to perform the various conditions in this study. This would provide a relative index of their experience of difficulty.

Performance Measures

A selection of dependent measures was used to capture the impact the different tasks have on performance (Table 1).

Table 1. The Performance Measures.

Variable - Source	Metric
Driving Behaviour – simulator data	Lane departures, standard deviation of lane position and root mean square error (RMSE) from lane centre. Mean speed, standard deviation of speed, RMSE speed. Standard deviation of following time headway, RMSE of time headway, minimum time headway. Reaction time to warning signs, missed warning signs and false alarms to warning signs.
Subjective Workload – rating scale for mental effort	Self-reported mental effort.
Conversation Quality– video/ audio	Errors, duration, pauses and failures.

Equipment

The TRL driving simulator was used in the proposed study (see previous). An after market hands-free phone kit was professionally fitted in the driving simulator. The same phone was used for both hands-free and hand-held conversations. A Nokia 3310 was used for this study based on evidence that theNokia 3310/30 range was the most popular/widely chosen handset on the UK market at the time of this study (February, 2002). The 3310 and 3330 are essentially the same phone. They look the same and have the same software. The only difference is that the 3330 has an extra menu option for Wireless Application Protocol (WAP) based services on the internet.

Questionnaires

The mobile phone questionnaire consisted of 28 questions designed to investigate all aspects of participant's mobile phone use (see Appendix C). The questionnaire is divided into four main sections. The first section is general questions looking at type of phone owned, typical type of phone conversations and frequency of phone use. The second section looked at phone user attitudes towards penalties for being caught whilst driving and using a mobile phone. The third section examined their own experiences of driving whilst using a mobile. The forth presented the participants with a list of tasks that people do typically while driving and they were asked to rate how distracting they thought each of these tasks would be on a Likert scale.

Task mental load was measured using a subjective rating scale: the Rating Scale Mental Effort (Zijlstra, 1993). Ratings of invested effort are indicated by a cross on a continuous 150 mm line (see Appendix C). Along the line, at several anchor points, statements related to invested effort are given, e.g., `almost no effort' or `extreme effort'. It is scored by measuring the distance from the origin to the mark in mm. With the Mental Effort Scale the amount of invested effort into the task has to be indicated, and not the more abstract aspects of mental workload (e.g., mental demand, as is in the NASA- TLX). These properties make the Mental Effort Scale ideal for self-report workload measurement (de Waard, 1996).

Analysis

Analysis of Driving Performance

The data from the simulator experiment was analysed to clarify the principal hazards of phone use while driving (e.g., poor vehicle control, slow reaction time). All of the data had a common time stamp allowing precise measures of performance across each of the tasks. Descriptive analyses were performed on all of the data from the experiment (i.e., central tendencies and distributions). The data was screened for anomalies (e.g., implausible values) and violations of parametric assumptions. This was followed by the inferential analyses. The principal analysis was a one-way repeated measures Analysis of Variance (ANOVA) to identify differences among the means of the four conditions (Control, Hand-held, Hands-free and Alcohol). Specific post-hoc comparisons were made where appropriate. All p-values were 2-tailed. The median scores of the three alcohol conditions were always used for the analyses. Graphical representations of the analyses, effect sizes and statistical significance levels were used to interpret the results.

Analysis of Conversations

Participants had three different tasks as part of their conversation, verbal puzzles, remembering sentences and monologues. They were asked to answer as accurately as possible and they were not to just guess the answers. Verbal puzzles were recorded as either correct or incorrect, the time that it took for them to respond was also recorded as a measure of the workload.

Sentences were scored on a scale of 0-3; zero if the participant could not remember the sentence that they had to repeat at all. They were awarded one point if they remembered a part of the sentence, twopoints if they remembered most of the sentence and were almost correct, and three points if they got the sentence 100% correct. The length of time it took for the participants to repeat back the sentence was also recorded along with the number of pauses that the participants made.

For the monologues, participants were given a topic which they had to talk about until they were asked to stop. Topics ranged from giving directions to describing a friend or relative. The number of words were recorded and divided by the length of the monologue to give the rate of words per second. Number of pauses and ums/ers were counted per monologue again as a measure of workload.

A tally of conversation failures was also recorded during this analysis. These were incidents where the driver completely failed to respond.

RESULTS

Driving Performance

Mean Speed

A one-way repeated measures ANOVA was calculated for mean speed on the motorway across the four conditions (Control, Hand-held, Hands-free and Alcohol). There was a significant main effect by condition for mean speed on the motorway [$F(3, 54) = 3.18$, $p < 0.05$, $\eta^2 = 0.15$]. On the motorway, drivers with hand-held phones drove significantly slower than during the Alcohol drive (p = 0.003). The other comparisons were not significantly different.

On the curves, where drivers were instructed to maintain a speed of 60mph, drivers drove closest to the specified limit in the control condition (see Figure 2). They exceeded the limit in the Alcohol drive. They went slower than the limit in the Hands-Free drive and they drove the slowest in the Hand-Held drive. There was a significant difference in speed across the conditions {$F(3, 54)$ = 10.41, $p < 0.001$, η^2 = 0.37}. When using hand-held phones, they drover significantly slower than during the Control ($p < 0.001$), Alcohol (p = 0.002) and Hands-free drive (p = 0.03).

The same pattern of mean speed behaviour was observed for the dual-carriageway section as was observed on the previous sections. There was a significant main effect by condition for mean speed on the motorway {$F(2.1, 38)$ = 9.69, $p < 0.001$, η^2 = 0.35}. There were problems of sphericity with the data so a Huynh-Feldt correction was used. When using hand-held phones, they drover significantly slower than during the Control ($p < 0.001$), Alcohol ($p < 0.001$) and Hands-free drive ($p < 0.001$).

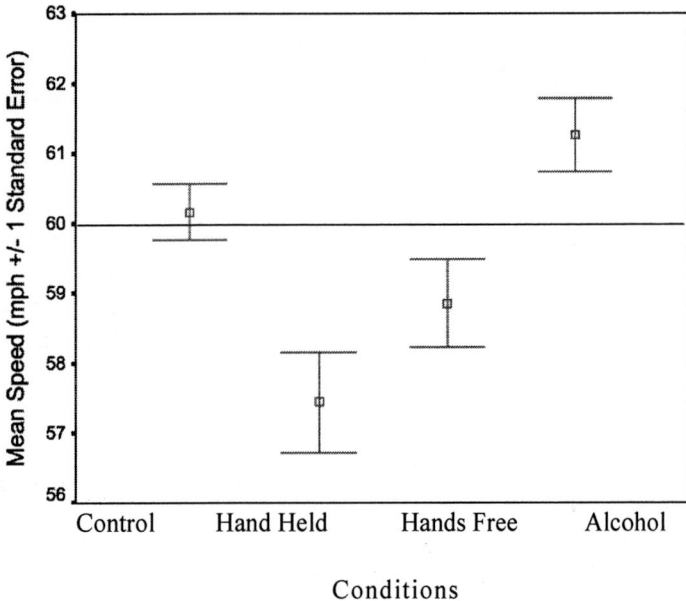

Figure 2. Mean Speed by Condition when Driving as Close to 60 mph as Possible.

Standard Deviation of Speed

There was no significant difference in the standard deviation of speed on the motorway drive. On the curves, there was a significant difference in standard deviation of speed across the conditions $\{F(3, 51) = 4.66, p < 0.01, \eta^2 = 0.22\}$. When using hand-held phones, speed was significantly more variable than during the Control ($p = 0.012$), Alcohol ($p = 0.001$) and Hands-free drive ($p = 0.014$).

Error from 60 mph on Curves

The root mean square error (RMSE) from 60 mph was calculated for the curved section of the route where drivers were instructed to maintain a speed as close to 60 mph as possible. The speed performance data was significantly skewed so a nonparametric Friedman's test was used to compare conditions. Speed keeping performance (Error from 60 mph on curves) was significantly different across the four conditions $\{$Chi-Square $= 17.8, p < 0.001\}$. Post-hoc comparisons were made using the Wilcoxon test. Speed keeping performance was significantly poorer in the Hand-Held condition in comparison to Normal driving $\{Z(19) = 2.80, p < 0.005\}$, Alcohol $\{Z(19) = 2.48, p < 0.01\}$ and Hands-free $\{Z(19) = 2.38, p < 0.05\}$. There were no significant differences among the other conditions.

Headway During Car Following

The amount of time the drivers were following the lead vehicle at a time-headway of less than 1 second was calculated. This measure was significantly skewed. A Friedman's test indicated no significant main effect for this measure. The RMSE for time and distance headway from the target following distance were also calculated. No significant differences were observed. Nothing was found for the standard deviation of time headway either.

Lane Keeping Performance

There were no differences among the chosen measures of lane keeping (standard deviation of lane keeping, lane departures, RMSE lane centre) for most of the route. The exception was there was a significant difference in the standard deviation of lane keeping on the dual-carriageway across the four conditions $\{F(1.6, 29) = 5.99, p < 0.01, \eta^2 = 0.25\}$. There were problems of sphericity with the data so a Huynh-Feldt correction was used. When under the influence of alcohol, lane keeping was significantly less steady than during the Control ($p = 0.05$), Hand-held ($p = 0.025$) and Hands-free drive ($p = 0.025$).

Reactions to Road Signs

There were 12 signs that appeared and 3 of these were target signs. Reaction times were calculated as the average response time of the hits. The road sign reaction time data was significantly skewed and kurtotic so a Friedman's test was used to compare conditions. Reaction time performance was significantly different across the four conditions {Chi-Square = 26.6, $p < 0.001$}. Post-hoc comparisons were made using the Wilcoxon test. Reaction times were significantly slower for the Hand-Held condition in comparison to normal driving {$Z(19) = 3.59, p < 0.001$} and Alcohol {$Z(19) = 3.29, p < 0.001$}. Hands-free were also significantly slower for the hand-held condition in comparison to normal driving {$Z(19) = 3.41, p < 0.001$} and alcohol {$Z(19) = 2.64, p < 0.01$}. The drivers were significantly slower when they had alcohol than in the control condition {$Z(19) = 2.56, p < 0.01$}. Hand-held was slower than hands-free but the difference was not significant, (see Figure 3).

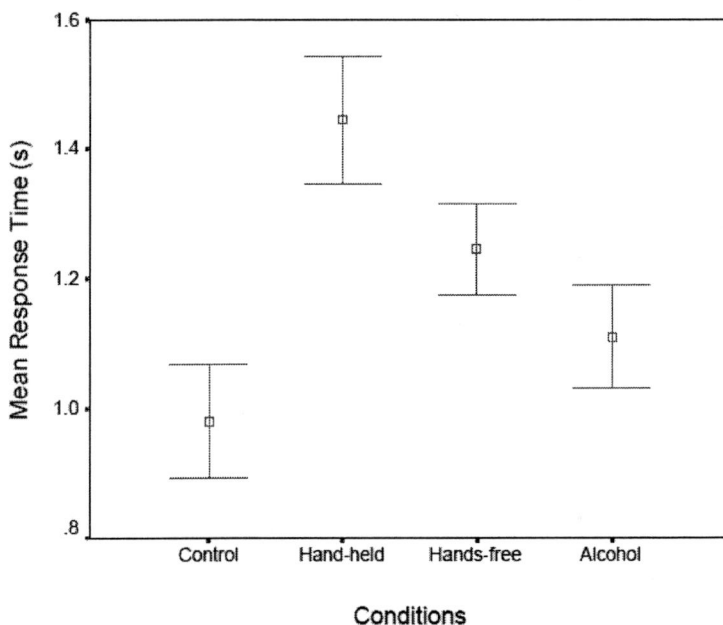

Figure 3. Mean Reaction Time to Warning Signs by Condition (± 1 Standard Error).

The misses and false alarms for the road signs were also analysed. Ideally the drivers should have had no misses or false alarms and three hits.

This response data was significantly skewed and kurtotic so a Friedman's test was used to compare conditions. The number of missed signs was significantly different across the four conditions {Chi-Square = 20.6, $p < 0.001$}. Post-hoc comparisons were made using the Wilcoxon test. There were significantly more missed for the hands-free {$Z(19) = 2.91, p < 0.005$} and hand-held {$Z(19) = 2.07, p < 0.05$} condition in comparison to normal driving. Hands-free also had significantly more misses than alcohol {$Z(19) = 2.86, p < 0.005$}. The number of false alarms signs was also significantly different across the four conditions {Chi-Square = 9.3, $p < 0.05$}. Post-hoc comparisons found significantly more false alarms for the hands-free {$Z(19) = 2.06, p < 0.05$} in comparison to normal driving. Hands-free {$Z(19) = 2.41, p < 0.05$} and hand-held {$Z(19) = 2.06, p < 0.005$} also had significantly more false alarms than alcohol.

There were no gender differences observed among the performance data across any of the route sections or treatment conditions.

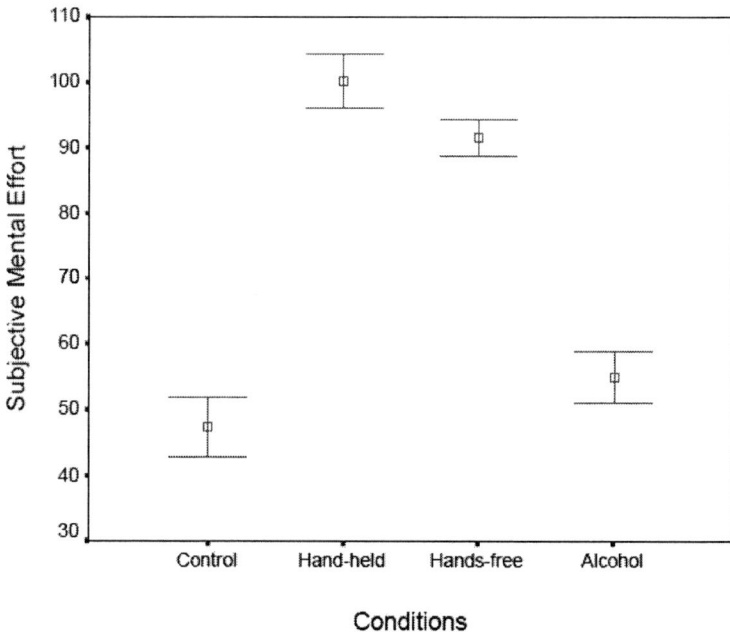

Figure 4. Mean Subjective Mental Effort Ratings by Condition (± 1 Standard Error).

Subjective Workload

A one-way repeated measures ANOVA was calculated for the subjective mental effort ratings across the four conditions (see Figure 4). There was a significant main effect by condition for mental effort $\{F(3, 57) = 66.62, p < 0.001, \eta^2 = 0.78\}$. Post hoc tests were run to compare the mean mental effort ratings by condition. Mental effort was rated highest for the hand-held drive and lowest for the Control drive. The Control required significantly less mental effort than the Hand-held ($p < 0.001$) and Hands-free ($p < 0.001$) conditions. The Alcohol was also rated significantly less demanding than the Hand-held ($p < 0.001$) and Hands-free ($p < 0.001$) conditions. Also, Hands-free required significantly more mental effort than Hand-held ($p = 0.008$). The Alcohol drive was rated more demanding than the Control drive, however this difference was not significant.

Conversation Performance

There were three conversation conditions (Hands-free, Hand-held and Control). The phone conversations were performed while driving and the conversations in the control condition were done while seated in a waiting area. Participants were asked how did the conversations compare in complexity to their normal phone conversation while driving? All drivers rated the conversations as being more difficult.

The rate of talking was calculated for the conversations by counting the words and dividing it by the duration of that segment of conversation. A one-way repeated measure ANOVA was calculated for the rate of talking across the three conversation conditions (see Figure 5). There was a significant main effect for condition $\{F(1.3, 25) = 45.44, p < 0.001, \eta^2 = 0.71\}$. There was a significant problem of sphericity with the data so a Huynh-Feldt correction was used. Post hoc tests were run to identify the differences. The rate of talking was significantly faster in the control conversations than in either the Hand-held ($p < 0.001$) or Hands-free conversations ($p < 0.001$). There were no differences between the two phone conditions.

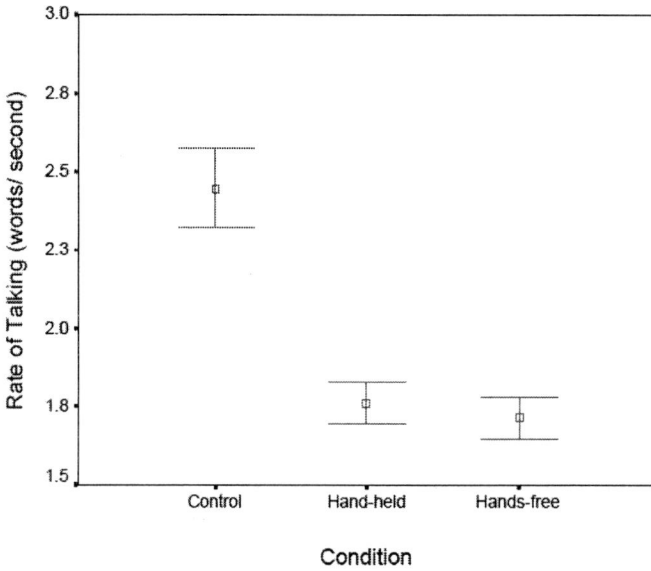

Figure 5. Mean Rate of Talking by Condition (± 1 Standard Error).

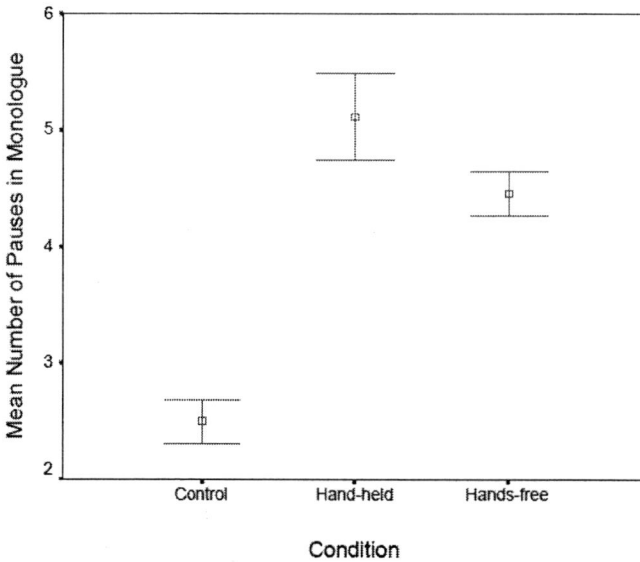

Figure 6. Mean Number of Pauses During the Monologues (± 1 Standard Error).

The number of pauses during the monologue portion of the conversations was calculated. A one-way repeated measure ANOVA was calculated for the number of pauses across the three conversation conditions (see Figure 6). There was a significant main effect for the number of pauses by condition $\{F(2, 38) = 35.31, p < 0.001, \eta^2 = 0.65\}$. Post hoc tests were run to compare the conditions. The number of pauses was significantly less in the control conversations than in either the Hand-held ($p < 0.001$) or Hands-free conversations ($p < 0.001$). There were also significantly fewer pauses when using the hands-free phones than when using hand-held ($p = 0.05$).

A one-way repeated measure ANOVA was also calculated for the number of pauses during the remembering sentences part of the conversations (see Figure 7). There was a significant main effect for the number of pauses by condition $\{F(2, 38) = 42.29, p < 0.001, \eta^2 = 0.69\}$. The number of pauses was significantly less in the control conversations than in either the Hand-held ($p < 0.001$) or Hands-free conversations ($p < 0.001$). Contrary to the monologues, there were also significantly fewer pauses when using the hand-held phones than when using hands-free ($p = 0.001$).

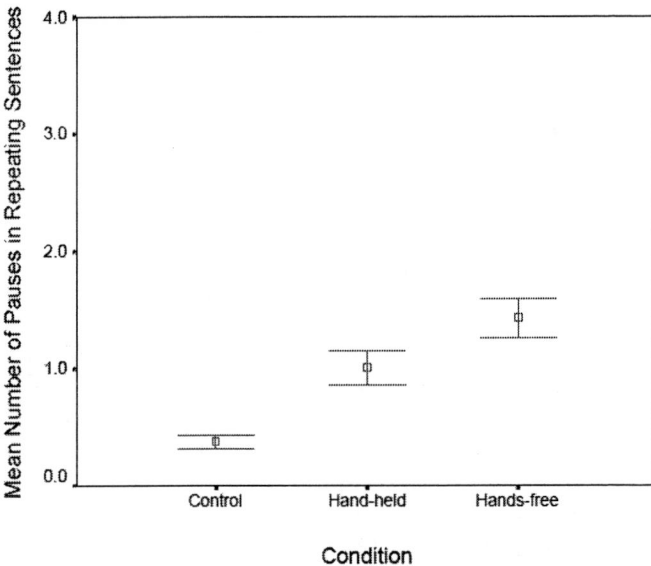

Figure 7. Mean Number of Pauses When Repeating Sentences (± 1 Standard Error).

The same analysis was done for the number of correct responses by condition (see Figure 8). There was a significant main effect for the number of correct answers by condition $\{F(2, 38) = 21.53, p < 0.001, \eta^2 = 0.53\}$. The number of correct answers was significantly greater in the control conversations than in either the Hand-held ($p < 0.001$) or Hands-free conversations ($p = 0.039$). There were also significantly more correct answers when using the hands-free phones than when using hand-held ($p = 0.001$).

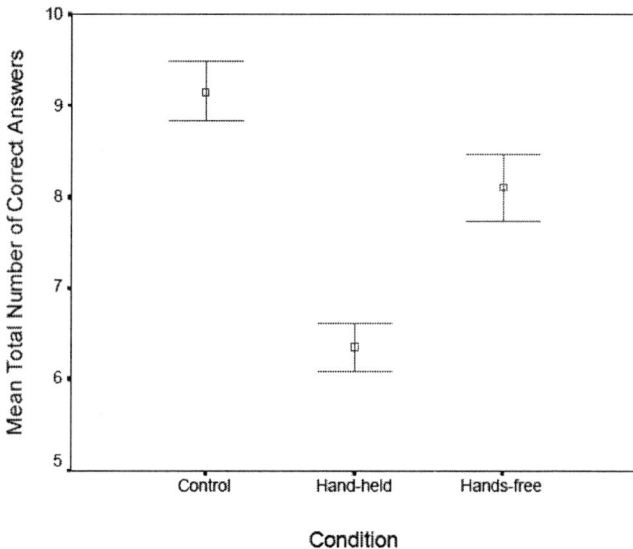

Figure 8. Mean Number of Questions Answered Correctly ($\pm$ 1 Standard Error).

A one-way repeated measure ANOVA was also calculated for the mean response time of repeating sentences (see Figure 9). There was a significant main effect for the time by condition $\{F(2, 38) = 111.75, p < 0.001, \eta^2 = 0.86\}$. The time to answer was significantly less in the control conversations than in either the Hand-held ($p < 0.001$) or Hands-free conversations ($p < 0.001$). The time was also significantly less when using the hand-held phones than when using hands-free ($p < 0.001$).

A one-way repeated measure ANOVA was also calculated for the mean response time of verbal puzzles (see Figure 10). There was a significant main effect for the time by condition $\{F(2, 36) = 6.93, p < 0.005, \eta^2 = 0.28\}$. The time to answer was significantly less in the control conversations than in either the Hand-held ($p = 0.002$) or Hands-free conversations ($p = 0.014$).

There was no significant difference in the response times between the phone conditions.

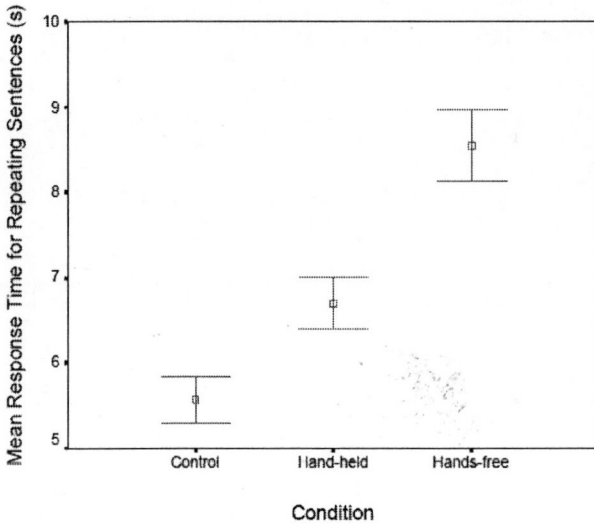

Figure 9. Mean Response Time for Repeating Sentences (± 1 Standard Error).

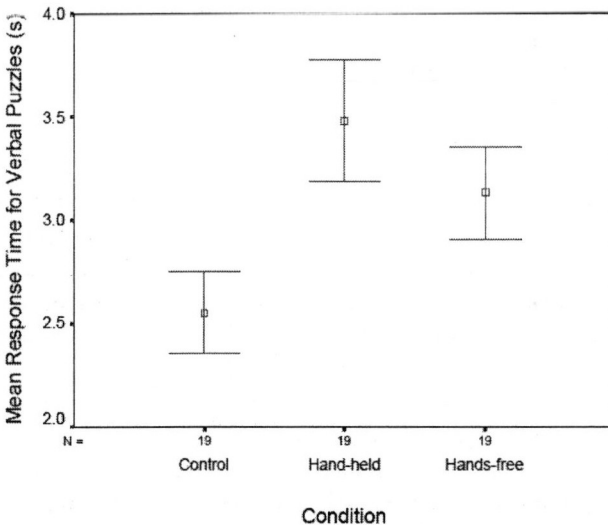

Figure 10. Mean Response Time for Verbal Puzzles (± 1 Standard Error).

The number of conversation failures was also recorded. These were incidents where the driver completely failed to respond. The data was skewed and kurtotic so a Wilcoxon test was used to compare the two conditions. Although there was twice as many failed conversations with the handheld (19) than with the hands-free phones (10), this difference was not statistically significant $\{Z(19) = 1.65, p = 0.098\}$.

Phone Use Questionnaire

All subjects owned their own mobile phone. Nokia was the most common brand (n = 8). Most respondents used their phones for personal reasons (n = 15) rather than business (n = 3). Two respondents only used their phones for emergencies. More than half (n = 14) used their phone every day or more often. Only seven respondents used hands-free equipment for their phones.

Five respondents said they would make phone calls while driving and seven would not. Others said it would depend on the caller or the traffic situation. Eight respondents said they would answer their phone if it rings while they were driving and four would not. Others said it would depend on the caller or the traffic situation. For incoming text messages, three respondents said they would read them while driving and eight said never. Others said it would depend on the caller or the traffic situation. Most respondents would never send a text message while driving (n = 16). One driver did send text messages and three said it depended on the traffic situation.

In terms of frequency, most respondents would not normally use their phones while driving (n = 12). Six respondents would often use their phone while driving. Respondents were also asked to estimate the duration of their typical carphone conversation. Estimates were between 1 and 2 minutes on average. People spoke on the phone for up to 80% of their driving time, with most only being on the phone for around 1% of driving time. The majority of respondents used their phone for short and simple conversations (n = 6) or to exchange brief messages (n = 9). It was believed that business people were most likely to use their phones while driving (n = 16).

In the questionnaire, respondents were asked to provide some opinions about using phones while driving. All but one respondent believed that using hand-held mobile phones while driving should be banned. A majority of respondents said there should not be a ban on hands-free phones (n = 11).

Only two thought there should also be a ban on hands-free phones. The remaining seven said it depends.

In terms of penalties, a majority of respondents said there should be a fine for using hands-held phones while driving (n = 11). Five believed that drivers should have points taken off their licence and 3 thought the penalty should be a written warning. Most respondents felt there should be not penalty for using a hands-free phone while driving (n = 13).

Most respondents thought that drivers should pull over at a safe place along their route and stop before using their phone (n = 16). All respondents would not use their phones during difficult driving conditions. These included bad weather, heavy traffic, motorways and in city centres. Some respondents stated that they would not use their phones when there was a police car in sight.

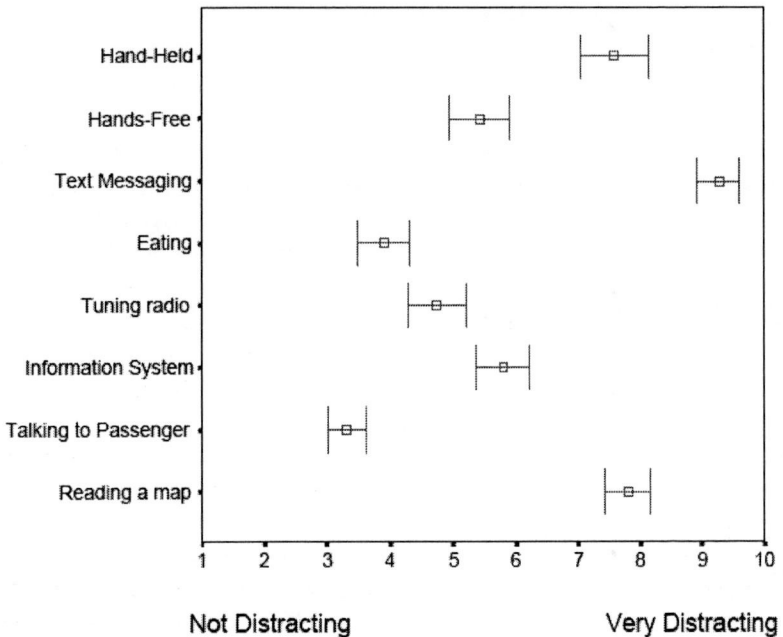

Figure 11. Mean Distraction Ratings (± 1 standard error).

A repeated measures ANOVA was calculated on the eight different driver distraction ratings (see Figure 11). There was a significant main effect for type of distraction $\{F(5, 93) = 29.10, p < 0.001, \eta^2 = 0.62\}$. There was a significant problem of sphericity with the data so a Huynh-Feldt correction was used. Post hoc tests were run to compare the distraction ratings by condition. Sending a text message was considered to be the most distracting activity to perform while driving. From the list, the least distracting activity was having a conversation with a passenger. The subjects rated talking with a passenger as being significantly less distracting than talking on a hands-free phone ($p < 0.001$). Tuning a radio was believed to be significantly less distracting than using a hand-held phone ($p = 0.001$) and reading a map ($p < 0.001$), but not a hands-free phone. Reading a map was rated as being more difficult than talking hands-free ($p = 0.002$). Talking hand-held was rated as being more difficult than talking hands-free ($p = 0.001$).

Summary of Results

Driving Performance

There was a tendency for drivers to slow down when talking on hand-held or hands-free phones, even when they were specifically instructed to maintain a set speed. Alcohol tended to have the opposite effect such that drivers drove faster than normal when under the influence of alcohol. The standard deviation of speed and speed error measures indicated that drivers' had significantly poorer speed control when using the hand-held phone than during the other three conditions. When drivers were under the influence of alcohol, they were significantly worse at driving smoothly (standard deviation of lane position) than during the other three conditions.

Towards the end of their route, drivers were instructed to flash their lights whenever a particular warning sign appeared at the roadside (e.g., pedestrians crossing). Reaction times were significantly slower for drivers using phones in comparison to when they had alcohol. Drivers were significantly slower when they had alcohol then when they had no distractions (i.e., control condition). Also, the drivers missed significantly more of these target signs when they were using a phone. There was also significantly fewer signs missed by the drivers when they were on alcohol in comparison to when they were using the hands-free phones. The phone drivers were also responding to the wrong warnings more often than the alcohol drivers (false alarms).

Subjective Ratings

From the subjective mental effort ratings participants made immediately after driving each route, it was clear that they found driving while using a hand-held phone to be the most difficult. The easiest task was the normal driving without any phone conversations. Hands-free was easier than hand-held. Drivers found it easier to drive drunk than to drive while using a phone, even when it was hands-free.

Conversation Performance

Conversations on the phones while driving were uniformly worse than conversations in the waiting room. Conversation performance comparisons between hand-held and hands-free phones were mixed. Hands-free phones were worse than hand-held phones for the repeating sentence tasks (time and number of pauses). Hand-held phones were worse than hands-free phones for the verbal puzzles (errors) and monologues (number of pauses). Although there was twice as many failed conversations with the hand-held than with the hands-free phones, this difference was not statistically significant. The conversations in this study were rated as being more difficult than the conversations participants would normally have on a phone while driving.

Phone Use Questionnaire

Sending a text message was considered to be the most distracting activity to perform while driving. From the list the least distracting activity was having a conversation with a passenger. The subjects rated talking with passengers as being significantly less distracting than talking hands-free. Tuning a radio was believed to be significantly less distracting than using a hand-held phone, but not a hands-free phone. Reading a map was rated as being more difficult than talking hands-free. Talking handheld was rated as being more difficult than talking hands-free.

Interpretations and Implications

Research has convincingly shown that phone conversations impair driving performance. Until now, it has been difficult to quantify the risk of this impairment because the comparison was to normal driving without using a phone. The aim of this study was to benchmark driving performance while using a mobile phone to a level of performance that is accepted by society as dangerous, i.e., driving with a blood alcohol level over the legal limit.

Results from this study showed a clear trend for significantly poorer driving performance (speed control, warning detection and response) when using a hand-held phone in comparison to the other conditions. The best performance was for normal driving without phone conversations. Hands-free was better than hand-held. Driving performance under the influence of alcohol was significantly worse than normal driving, yet tended to be better than driving while using a phone. Drivers also reported that it was easier to drive drunk than to drive while using a phone. The exception was with lateral control, where alcohol impaired lane-keeping performance more than the phones.

Alcohol is a central nervous system depressant that impairs the ability to perform complex tasks. Of relevance to driving, alcohol impairs the skills associated perception, divided attention, attentional shift, working memory, motor co-ordination, reaction time and rate of information processing (e.g., Moskowitz & Burns, 1990). The consequence of this impairment is that drivers with a 0.08% BAC have a four to five times greater risk of being involved in crash (e.g., Vinson, Mabe & Leonard, 1995). From the results of the present study, it could be argued that drivers would be exposed to an equivalent risk while using a phone.

The critical finding in this study was that phone use impaired drivers' abilities to respond to warnings more so than alcohol. As Mcknight and Mcknight (1991) point out, the number of missed warnings has more significance for safety than the slowing of response times. The road traffic environment may have some tolerance for a delay in responding to a hazard. There is no tolerance for missed hazards.

The effect of phone use on speed has been observed in previous research on driver distraction (e.g., Alm & Nilsson, 1994). In fact it appears to be one of the more consistent finding in the research on driving with phones. One explanation is that drivers slow the vehicle in order to make the driving task easier. It is essentially a strategy they use to cope with the competing demands of multiple tasks. Another explanation is that drivers fail to monitor their speed and are unable to maintain their normal speed.

Although performance when using hands-free phones was worse than hand-held phones, hands-free phones still impaired reaction times more than alcohol. Thus, even though hands-free may be slightly less dangerous than hand-held phones, the safest approach would be to turn phones off while driving.

Researchers have attempted economic rationalism to defend the risk of death and injury from using mobile phones while driving (Lissy et al., 2000). Baring the ethical issues of accepting the loss of human life and health, this work is flawed because it does not fully consider the context of phone use in cars. Of course there are benefits to having mobile phones in cars. However the major benefits are not lost if the phone is turned off and a messaging service is used. In the event of a crash or a breakdown, the phone is still available to call emergency services.

Strengths and Weaknesses

It is important to discuss the potential problems of comparing the impairment in driving performance caused by alcohol to the impairment caused by distraction from a phone conversation. The motivation for this comparison was the need to provide a benchmark of driving performance while using a mobile phone to a proven danger. However, alcohol may impair driving performance differently than driver distraction. As described above, alcohol is a central nervous system depressant whereas using a phone while driving divides drivers' attentional resources. Divided attention through timesharing or multitasking tends to impair performance on one or more of the separate tasks (Wickens, 1992). The present study found that performance on both the conversation and driving tasks was impaired. During the alcohol condition, the drivers had no distractions and could concentrate fully on the driving tasks.

Also, the driver may have some control over their divided attention. A phone user can pause their conversation in order to deal with a change in the traffic situation (e.g., to watch for a changing traffic light). Once a drunk driver is behind the wheel there may not be that much he or she can do to control their performance. Drivers will also stay drunk for the duration of their drive while the distraction from phones may only occur for a small portion of driving. Thus, on a trip by trip basis, drunk driving would be much more dangerous because the exposure to risk is greater. This risk would change when looking at a population level. At any given moment the amount of drivers using their phones would greatly exceed the number drunk drivers.

It is difficult to draw any straightforward conclusions about crash risk and severity from the present study. This is because the situations when drunken driving and phone use occur may be different. Drink driving tends to occur late at night on weekends whereas phone use can happen at any time. For example, drivers can use their phones when driving in town centres or on motorways during busy rush hour traffic. This difference would lead to differences in crash type and severity. Alcohol crashes are more likely to involve younger male drivers in a single vehicle road departure late at night and result in serious injury to the occupants (e.g., Johnston, 1982; Maycock, 1997). Although there is insufficient data on mobile phone crashes, one might expect them to involve lower speeds, other road users and less severe injuries. Another issue with alcohol is that it impairs driver judgement. For example, a drunk driver might think they are driving well and that they are capable of driving faster. Phone use might delay or remove judgement, but it does not impair it as such.

One criticism of this work might be that only a small sample of drivers was used. The sample consisted of phone users and experienced drivers. It was not a sample of convenience and they were representative of the population of English drivers that use phones. If anything, the small sample size should emphasise the magnitude of the results. Statistically significant differences were found among the conditions despite the low statistical power of the inferential tests. Furthermore, the size of effect tended to indicate moderate to large differences in performance among the conditions.

There may be some concern that the phone conversations in this study were harder than typical in-car conversations. The material was difficult and the conversations, although not continuous, lasted for the duration of the test drive. However, this study made it much easier for drivers by only considering the distraction from talking, thinking and listening. No other phone tasks were examined. The phone was answered before driving and was hung-up when the trial was completed. The task of using a mobile phone in a vehicle has many task elements that vary in the amount and way they distract the driver. Detailed hierarchical task analyses of hand-held and hands-free mobile phones have been conducted (Kersloot & Lansdown, 1999). Some of these task elements would be more or less distracting. Fuse et al (2001), in a small simulator study, identified that driver reaction times were slower when making and receiving calls than during the conversation. They supported these findings with Japanese police reported crash data from 1997 that found the majority (69%) of crashes caused by phones were attributed to making and receiving calls. A much smaller amount occurred during the phone conversations (16%). Also, this only includes the direct task elements and does not include other distractions associated with making a call. Callers also need to plan the call, remember numbers, take notes, find information to support the conversation (e.g., a calendar), and reflect on the conversation afterwards (de Waard, et al, 2001). Furthermore, there is the distraction cause by the interference and interruptions of poor phone connections. Other functions offered by phones might also be a dangerous distraction, for example reading and composing text messages. Therefore, it is more likely that the phone tasks in this study were less complex than 'real' use. Consequently, driving performance when making and receiving calls can be expected to be even more impaired.

Future Research

Nearly all road safety research (99%) is based on crash records that do not contain essential information about manoeuvres, immediate circumstances, indirectly involved road users and road features (Oppe, 2002). Because of this lack of fundamental information, we are unable to determine the contribution of such things as mobile phones to causing crashes. It is essential that crash data include information on the presence of mobile phones in the vehicle. With improved accident reporting, the effects of mobile telephones on crash involvement can be investigated and the precise hazards of phone use in cars will become even clearer.

Research is also needed to replicate the results of the present study in other countries and on a larger sample of drivers.

The Mobile Telecommunications Research Programme (MTHR) has funded TRL to conduct a study to compare the hazards of hands-free conversations with more conventional in-vehicle distractions like talking with passengers and operating the radio. This MTHR research is currently underway and the results will be available early in 2003.

CONCLUSIONS

Driving while intoxicated is clearly dangerous and this study further confirmed that alcohol impairs driving performance. However, this study also found that certain aspects of driving performance are impaired more by using a phone than by having a blood alcohol level at the legal limit (80mg/100ml). It is concluded that driving behaviour while talking on a phone is not only worse than normal driving, it can also be described dangerous.

Although it is dangerous, illegal and irresponsible to use a hand-held phone while driving, this behaviour is common. Phone users should be discouraged more from engaging in this dangerous behaviour while behind the wheel. It is hoped that this research will contribute to the evidence for convincing people to turn their phones off while driving.

Acknowledgements

The authors would like to thank Direct Line Insurance for commissioning this research.

APPENDIX A: INSTRUCTIONS

PLEASE ANSWER IN HANDHELD MODE THE MOBILE PHONE THAT IS RINGING BESIDE YOU. YOUR INSTRUCTIONS FOR THIS TRIAL SESSION AND SOME VERBAL TASKS WILL BE RELAYED THROUGH THIS PHONE

REACTION TIME TASK
Please look at image on the seat beside you: CYCLIST at stages during the course of your drive various warning triangle signs will appear. For this drive you are looking for the CYCLIST sign only. As soon as you see it please respond by flashing the headlights, so that we can measure your reaction time. Please practice now.

FOLLOWING TASK
Please study the distance between you and the vehicle ahead. As this vehicle travels along the motorway it will vary its speed. Please start the engine, and follow it maintaining the current DISTANCE, no more no less.

NORMAL TRAFFIC
Your following exercise is over. Now please drive as you would normally on the motorway.

LINK - END MOTORWAY
Please move over to the inside lane.

CURVES
As you go under the bridge I would like you to be travelling at 60mph. Please maintain a speed of 60mph and at the same time keep a line as close to the centre of this left-hand lane as you can.

CHOICE REACTION

END

Thank you that is the end of your session. Please stop the vehicle and then apply the handbrake, take it out of gear and switch the engine off.
Please complete the mental effort scale in your booklet.

APPENDIX B: CONVERSATION TASKS

Conversation A

RS – Repeat sentence
VP – Verbal puzzle
M - Monologue

RS1   Felix is darker than Antoine. Who is the lighter of the two?
VP1   The action of the brave cyclist kept the small boy from being hit by the 10-ton truck.
M1   YOUR SITTING ROOM
VP2   If you see a circle and it has a rectangle to the right of it and if there is a cross directly below the rectangle. Is the rectangle:

a. Below the cross?
b. To the left of the Circle?
c. Below the circle

RS2   It was raining this morning so the children wore their boots to school.
VP3   If Daphne walks twice as fast as Margaret and they are the only two people in a race, who is most likely to finish last?
RS3   Annie's dog ran to her for help after it was attacked by a racoon in the woods.
M2   A MEMORABLE HOLIDAY
VP4   If three chocolate bars cost 93 pence, what is the cost of one chocolate bar?
RS4   The team was playing well until the third quarter, when snow made visibility poor.
VP5   Horse number seven entered the home stretch before Tom, number eight's jockey, could get his horse out of the gate.

a. Which horse was Tom riding?
b. Where was horse seven?

RS5   Police protection was given to Mary after her apartment was broken
       into by a daring thief.
M3   A FRIEND
VP6   Jack, who was working in Tim's garage, found an old MG that
       belonged to his father.

a. Who did the car belong to?
b. Where was Jack?

RS6   The car lost power trying to accelerate on the slippery hill during a
       storm in March.
VP7   If a car drove 360 miles in six hours, how fast was the car going in
       miles per hour
RS7   The train crept up the mountain slowly as it wended its way through
       the Rockies.
M4   ROUTE TRL TO BRACKNELL
VP8   Who is sicker if Jane is less ill than Sam?
RS8   Jane started dancing at age eight, but didn't give her first recital until
       she was twenty-three.
VP9   If you see a picture with a cross beneath a rectangle, but to the right of
       a circle, is the rectangle:

a. Above the circle?
b. To the left of the circle?
c. Right of the cross?

RS9   The perfume was strong, but Jane liked the exotic scent of Jasmine.
M5   ANIMALS BEGINNING B

Conversations B

VP1   If I say Jack stole Ann's ball who is the thief?
RS1   The driver was stopped for driving 67 miles per hour in a 20 mile per
       hour zone.
M1   THE INTERIOR OF YOUR CAR
VP2   If you see a picture with a diamond, a rectangle and a circle, and the
       circle is to the right of the rectangle and directly above the diamond,
       is the rectangle:

a. Right of the diamond?
b. Above the circle?
c. Left of the circle?

RS2   Undetected by the Sleeping dog, the thief broke into Jane's apartment.
VP3   Which girl is taller if Jane is shorter than Kim?
RS3   Mike walked around the block three times before he had the nerve to knock on Carol's door
M2   YOUR DAILY WORK / ACTIVITIES
VP4   If Jane runs 6 miles in 54 minutes, how long does it take her to run one mile?
RS4   The train left Cleveland an hour early, leaving Sam stranded at the station
VP5   The man who was an engineer came to the store where Alice worked to buy pastries.

a. Who bought the pastries?
b. Where was Alice

RS5   The shorter the chapter, the easier it is for students to complete the difficult exercises.
M3   A PAST / PRESENT BOSS
VP6   Because he was working late, Jack left a dinner in his microwave for Jim to heat up when he got home.

a. Who was the dinner for?
b. Who did the Microwave belong to?

RS6   The warm humid weather that occurs in the tropics makes people sleepy by midday
VP7   A chocolate bar costs 24 pence. What will 3 chocolate bars cost?
RS7   Old houses are more difficult to maintain, but worth the extra time and effort.
M4   ROUTE FROM YOUR HOME TO M3 OR M4 MOTORWAY
VP8   Which house is smaller if Jim's house is half again as big as Brian's?
RS8   The students needed to complete chapters 9 and 11 and answer the question on page twenty.
VP9   If you see a picture with a circle to the left of a square but on top of a cross, is the cross:

a. Above the square
b. To the left of the circle?
c. Below the circle?

RS9 The weather in March is snowy and cold in many parts of Canada.
M5 NAMES BEGINNING A

Conversations C

VP1 If I say Jane is wearing Alison's coat, who does the coat belong to?
RS1 The boat developed engine problems as it left port, leaving passengers wondering how long they would be delayed.
M1 YOUR LOCAL SUPERMARKET
VP2 If you see a picture with a circle and the cross is to the left of it and a rectangle is directly above the cross, is the cross:

a. Below the rectangle
b. Left of the rectangle?
c. Right of the circle?

RS2 Chased by an angry cat, the mouse burrowed deeply into the woodpile.
VP3 If Charles beats David in a sprint, which man is the faster running? Charles
RS3 If you have occasion to visit the tropics try and go when ocean breezes make sailing fun.
M2 LAST CHRISTMAS
VP4 If one pair of Jeans cost £21, how much will fours pairs cost? £84
RS4 Due to foresight and planning the family was able to realize their dream vacation.
VP5 Janice, the head librarian, walked to the seventh floor, where John was shelving books.

a. Where was John?
b. What did Janice do?

RS5 The video camera captured the bank robber's daring daylight robbery of the First Avenue Bank.
M3 A RELATIVE

VP6   In the backyard of Joe's house, Alice and Frank's dog played Frisbee

a. Who did the dog belong to?
b. Where were they playing?

RS6   The car was clearly out of control as it careened across the median
      and into ongoing traffic.
VP7   How many hours will it take to run 21 miles at a rate of three miles
      per hour?
RS7   Because he ripped his shirt on the nail, Sam had to mend the pocket.
M4   ROUTE TRL TO WOKINGHAM
VP8   Eric is one and a half times as big as Allen. Who is the smaller?
RS8   The old house had cedar shingles and the floor sagged from five
      generations of scrambling children.
VP9   If you see a picture with a diamond to the right of a circle and a
      square below the circle, is the circle:

a. Above the square?
b. Below the diamond?
c. Left of the square?

RS9   The wild flowers bloomed in profusion in the high meadows in
      August.
M5   TOWNS BEGINNING R

APPENDIX C: SAMPLE SUBJECT CASE REPORT FORM

See Separate File: Sample Case Report Form

REFERENCES

Alm, H. & Nilsson, L. (1994). Changes in driver behaviour as a function of
    hands-free mobile phones. *Accident Analysis and Prevention, 26*, 441-
    451.
Borkenstein, R.F., Crowther, F.R., Shumate, R.P., Ziel, W.B. & Zylman, R.
    (1964). The role of the drinking driver in traffic accidents. Department
    of Police Administration, Indiana University. Brookhuis K A, de Vries

G and de Waard D (1991). The effects of mobile telephoning on driving performance. *Accident Analysis and Prevention*, **23**, 309.

Brown, I. D., Tickner, A. H. and Simmonds, D. C. (1969). Interference between concurrent tasks of driving and telephoning. *Journal of Applied Psychology*, **53**, 419-424. de Waard, D. (1996). *The measurement of drivers' mental workload*. PhD thesis, University of Groningen. Haren, The Netherlands: University of Groningen, Traffic Research Centre. de Waard, D., Brookhuis, K. and Hernandez-Gress, N. (2001). The feasibility of detecting phone-use related driver distraction, *International Journal of Vehicle Design*, **26** (1), 85-95. Fairclough S H, M Ashby , T Ross and A M Parkes (1991). Effects of hands-free telephone use on driving behaviour. *Proceedings of the ISATA International Symposium on Automotive Technology and Automation*, Florence, May, 403-409.

Cognitive workload while driving and talking on a cellular phone or to a passenger. *International Ergonomics Association Conference*, San Diego, USA.

Fairclough, S. H. and Graham, R. (1999). Impairment of driving performance caused by sleep deprivation or alcohol: a comparative study. *Human Factors*, **41** (1), 118-128.

Fuse, T., Matsunaga, K., Shidoji, K., Matsuki, Y. and Umezaki, K. (2001). The cause of traffic accidents when drivers use car phones and the functional requirements of car phones for safe driving, *International Journal of Vehicle Design*, **26** (1), 48-56. Goodman, M. J., Bents, F. D. Tijerina, L., Wierwille, W. Lerner, N. and Benel, D. (1997). *An investigation of the safety implications of wireless communications in vehicles*. United States Department of Transportation, National Highway Traffic Safety Administration, Washington, D.C. Johnston, I. R. (1982). The role of alcohol in road crashes. *Ergonomics*, **25** (10), 941-946.

Kersloot, T. and Lansdown, T. C. (1999) In-Vehicle Telephone Task Analysis. Presented at the *Intelligent Transport Systems (ITS) World Congress*, Toronto.

Lamble D, Kauranen T, Laakso M and Summala H (1999). Cognitive load and detection in thresholds in car following situations: safety implications for using mobile (cellular) telephones while driving. *Accident Analysis and Prevention*, **31**, 617.

Lissy K.S., Cohen I.T., Park M.Y., Graham J. (2000). *Cellular phone use while driving: Risks and benefits*. Harvard Center for Risk Analysis, Harvard School of Public Health: Boston, MA.

Maycock, G. (1997). *Drinking and driving in Great Britain – A review*. TRL Report 232. Crowthorne, U.K.

McKnight, A. J., and McKnight, A. S. (1991). *The effect of cellular phone use upon driver attention*. Washington, DC: National Public Services Research Institute. Funded by AAA Foundation for Traffic Safety.

Moskowitz, H., and Burns, M. (1990). Effects of Alcohol on Driving Performance. *Alcohol Health & Research World*, 14(1). 12-14.

National Police Agency Traffic Bureau (1998). *Annual Traffic Facts for 1997*. Japan.

Oppe, S. (2002). Siem Oppe Colloquium, *SWOV Institute for Road Safety Research Newsletter*, www.swov.nl, June 2002.

Parkes, A. (1991). The effects of hands-free telephone use on conversation structure and style. Paper presented at *the 24th Annual Conference of the Human Factors Association of Canada*, HFAC/ACE, Vancouver, BC. 141-146.

Redelmeir D. A. and Tibshirani, R. J. (1997). Association between cellular-telephone calls and motor vehicle collisions. *New England Journal of Medicine*, **336** (7), 453-458. Stewart Report (2000). Mobile Phones and Health, The Independent Expert Group on Mobile Phones, Sir W. Stewart (Chairman), (http://www.iegmp.org.uk).

Stevens, A. and Minton, R. (2001). In-Vehicle Distraction and Fatal Accidents in England and Wales. *Accident Analysis and Prevention*. **33**, 539-545.

Stevens, A. and Paulo, D. (1999). The use of mobile phones while driving. *TRL Research Report* **318**. Watson, P. E., Watson, I. D. and Batt, R. D. (1980). Prediction of blood alcohol concentration in human subjects. *Journal of Alcohol Studies*, **40**(7), pp. 547-556.

Vinson, D.C., Mabe, N. and Leonard, L.L. (1995). Alcohol and injury: A case-crossover study. *Archives of Family Medicine*, **4**, 505-11. Waugh, J. D., Glumm, M. M., Kilduff, P. W., Tauson, R. A., Smyth, C. C. and Pillalamarri, R. S. (2000).

Wickens, C. D. (1992). *Engineering Psychology and Human Performance, 2nd Edition*. Harper-Collins: New York.

Zijlstra, F.R.H. (1993). *Efficiency in work behaviour. A design approach for modem tools*, PhD Thesis Delft University of Technology, The Netherlands.

In: Mobile Phones and Driving   ISBN: 1-60021-162-3
Editor: D. M. Sturnquist, pp. 65-75   © 2006 Nova Science Publishers, Inc.

Chapter 3

REFERENCES

This bibliography, prepared by the National Safety Council Library[*] from materials in its collection, deals with research studies concerning cell phones and driving. It lists documents published from 1997 to 2005. Some documents have a URL listed, so you can freely download them. Otherwise, the Library can provide assistance in obtaining documents.

S 05 - 1157
MOBILE PHONE USE -- EFFECTS OF HANDHELD AND HANDSFREE
    PHONES ON DRIVING PERFORMANCE
Author(s): TORNROS JE / BOLLING AK
Source: ACCIDENT ANALYSIS & PREVENTION V.37 NO.5, PP.902-
    909, SEP 2005

S 05 - 1090
CELL PHONE USE ON THE ROADS IN 2002
Author(s): GLASSBRENNER D
Source: NATIONAL HIGHWAY TRAFFIC SAFETY ADMIN.,
    WASHINGTON, DC, 46PP. SEP 2005
URL:http://www-nrd.nhtsa.dot.gov/pdf/nrd-30/NCSA/Rpts/2005/809580.pdf

S 05 - 1050
IN-CAR CELL PHONE USE AND HAZARDS FOLLOWING HANDS
    FREE LEGISLATION
Author(s): RAJALIN S / SUMMALA H / POYSTI L

---

[*] E-mail address: library@nsc.org. The Library can be contacted at: Phone: 630-775-2199

Source: TRAFFIC INJURY PREVENTION V.6 ISSUE 3, PP.225-229, SEP
   2005

S 05 - 0956
EFFECTS ON WASHINGTON, D.C. LAW ON DRIVERS' HAND-HELD
   CELL PHONE USE
Author(s): MCCARTT AT / HELLINGA LA / GEARY LL
Source: INSURANCE INST. FOR HIGHWAY SAFETY, ARLINGTON,
   VA, 10PP. APR 2005

S 05 - 0781
ROLE OF MOBILE PHONES IN MOTOR VEHICLE CRASHES
   RESULTING   IN   HOSPITAL   ATTENDANCE: A   CASE-
   CROSSOVER STUDY
Author(s): MCEVOY SP / STEVENSON MR / MCCARTT AT
Source: BRITISH MEDICAL JOURNAL, 5PP. 12 JUL 2005

S 05 - 0682
HOW DANGEROUS IS DRIVING WITH A MOBILE PHONE?
   BENCHMARKING THE IMPAIRMENT TO ALCOHOL
Author(s): BURNS PC / PARKES A / BURTON S
Source: TRANSPORT RESEARCH LAB., CROWTHORNE, ENGLAND,
   33PP. 2005
URL: http://www.cartalk.com/content/features/Drive-Now/trl-study.pdf

S 05 - 0453
FACTORS INFLUENCING THE USE OF CELLULAR (MOBILE)
   PHONE DURING DRIVING AND HAZARDS WHILE USING IT
Author(s): POYSTI L / RAJALIN S / SUMMAIA H
Source: ACCIDENT ANALYSIS & PREVENTION V.37 NO.1, PP.47-51,
   JAN 2005

S 05 - 1016
CELL PHONES AND DRIVING PERFORMANCE: A META-ANALYSIS
Author(s): HORREY WJ / WICKENS CD
Source: PROCEEDINGS OF THE HUMAN FACTORS & ERGONOMICS
   SOCIETY, 48TH ANNUAL MEETING, SEPTEMBER 20-24, 2004,
   NEW ORLEANS, LOUISIANA (CD-ROM) PP.2304-2308, 2004

S 05 - 1012
WHAT DO DRIVERS FAIL TO SEE WHEN CONVERSING ON A CELL PHONE?
Author(s): STRAYER DL / COOPER JM / DREWS FA
Source: PROCEEDINGS OF THE HUMAN FACTORS & ERGONOMICS SOCIETY, 48TH ANNUAL MEETING, SEPTEMBER 20-24, 2004, NEW ORLEANS, LOUISIANA (CD-ROM) PP.2213-2217, 2004

S 05 - 1011
PASSENGER AND CELL-PHONE CONVERSATIONS IN SIMULATED DRIVING
Author(s): DREWS FA / PASUPATHI M / STRAYER DL
Source: PROCEEDINGS OF THE HUMAN FACTORS & ERGONOMICS SOCIETY, 48TH ANNUAL MEETING, SEPTEMBER 20-24, 2004, NEW ORLEANS, LOUISIANA (CD-ROM) PP.2210-2212, 2004

S 05 - 0629
RISK PERCEPTIONS OF MOBILE PHONE USE WHILE DRIVING
Author(s): WHITE MP / EISER JR / HARRIS PR
Source: RISK ANALYSIS V.24 NO.2, PP.323-334, 2004

S 05 - 0511
THE IMPACT OF IN-VEHICLE CELL-PHONE USE ON ACCIDENTS OR NEAR-ACCIDENTS AMONG COLLEGE STUDENTS
Author(s): SEO DC / TORABI MR
Source: JOURNAL OF AMERICAN COLLEGE HEALTH V.53 NO.3, PP.101-107, NOV/DEC 2004

S 04 - 0284
A COMPARISON OF THE CELL PHONE DRIVER AND THE DRUNK DRIVER
Author(s): STRAYER DL / DREWS FA / CROUCH DJ
Source: AEI-BROOKINGS JOINT CENTER FOR REGULATORY STUDIES, WASHINGTON, DC, 19PP. JUL 2004

S 05 - 0271
PROFILES IN DRIVER DISTRACTION: EFFECTS OF CELL PHONE CONVERSATIONS ON YOUNGER AND OLDER DRIVERS
Author(s): STRAYER DL / DREWS FA
Source: HUMAN FACTORS V.46 NO.4, PP.640-649, WINTER 2004

S 05 - 0268
EFFECTS OF PRACTICE ON INTERFERENCE FROM AN AUDITORY
TASK WHILE DRIVING: A SIMULATION STUDY
Author(s): SHINAR D / TRACTINSKY N
Source: NATL. HIGHWAY TRAFFIC SAFETY ADMIN.,
WASHINGTON, DC, 51PP. DEC 2004

S 05 - 0167
EFFECTS OF NATURALISTIC CELL PHONE CONVERSATIONS ON
DRIVING PERFORMANCE
Author(s): RAKAUSKAS ME / GUGERTY LJ / WARD NJ
Source: JOURNAL OF SAFETY RESEARCH V.35 NO.4, PP.453-464,
2004

S 04 - 1082
THE IMPACT OF DRIVER CELL PHONE USE ON ACCIDENTS
Author(s): HAHN RW / PRIEGER JE
Source: AEI-BROOKINGS JOINT CENTER FOR REGULATORY
STUDIES, WASHINGTON, DC, 80PP. JUL 2004

S 04 - 0631
CELL PHONE USE
Author(s): MONTERESSI C
Source: EXXONMOBIL BIOMEDICAL SCIENCES INC., ANNANDALE,
NJ, 20PP, 2004
URL: http://www2.exxonmobil.com/corporate/files/corporate/
Cell_Phone_Use_Study.pdf

S 04 - 0942
DRIVING PERFORMANCE DURING CONCURRENT CELL-PHONE
USE: ARE DRIVERS AWARE OF THEIR PERFORMANCE
DECREMENTS?
Author(s): LESCH MF / HANCOCK PA
Source: ACCIDENT ANALYSIS & PREVENTION V.36 NO.3, PP.471-
480, MAY 2004

S 04 - 0930
USING MOBILE TELEPHONES: COGNITIVE WORKLOAD AND
ATTENTION RESOURCE ALLOCATION
Author(s): PATTEN CJ / KIRCHER A / OSTLUND J

Source: ACCIDENT ANALYSIS & PREVENTION V.36 NO.3, PP.341-350, MAY 2004

S 04 - 0772
LONGER TERM EFFECTS OF NEW YORK STATE'S LAW ON DRIVERS' HANDHELD CELL PHONE USE
Author(s): MCCARTT AT / GEARY LL
Source: INJURY PREVENTION V.10 NO.1, PP.11-15, FEB 2004

S 04 - 0692
LIVING DANGEROUSLY: DRIVER DISTRACTION AT HIGH SPEED
Author(s): JOHNSON MB / VOAS RB / LACEY JH
Source: TRAFFIC INJURY PREVENTION V.5 ISSUE 1, PP.1-7, MAR 2004

S 04 - 0632
RISKY DRIVING: THE RELATIONSHIP BETWEEN CELLULAR PHONE AND SAFETY BELT USE
Author(s): EBY DW / KOSTYNIUK LP / VIVODA JM
Source: UNIV. OF MICHIGAN, ANN ARBOR. TRANSPORTATION RESEARCH INST. 12PP. 2003
URL: http://www.ltrc.lsu.edu/TRB_82/TRB2003-000807.pdf

S 04 - 0223
MOBILE PHONE USE WHILE DRIVING: CONCLUSIONS FROM FOUR INVESTIGATIONS
Author(s): THULIN H / GUSTAFSSON S
Source: SWEDISH NATL. ROAD & TRANSPORT RESEARCH INST., LINKOPING, SWEDEN, 36PP. 2004

S 04 - 0284
FATAL DISTRACTION? A COMPARISON OF THE CELL-PHONE DRIVER AND THE DRUNK DRIVER
Author(s): STRAYER DL / DREWS FA / CROUCH DJ
Source: PROCEEDINGS OF THE 2ND INTERNATIONAL DRIVING SYMPOSIUM ON HUMAN FACTORS IN DRIVER ASSESSMENT, TRAINING AND VEHICLE DESIGN, JULY 21-24, 2003, PARK CITY, UTAH, PP.25-30, JUL 2003
URL: http://www.psych.utah.edu/AppliedCognitionLab/ DrivingAssessment2003.pdf

S 04 - 0270
INVESTIGATING THE RELATIONSHIP BETWEEN CELLULAR
PHONE USE AND TRAFFIC SAFETY
Author(s): ABDEL-ATY M
Source: ITE JOURNAL V.73 NO.10, PP.38-42, OCT 2003

S 04 - 0248
DRIVER HAND-HELD MOBILE PHONE USE AND SAFETY BELT
USE
Author(s): EBY DW / VIVODA JM
Source: ACCIDENT ANALYSIS & PREVENTION V.35 NO.6, PP.893-
895, NOV 2003

S 04 - 0060
MOBILE TELEPHONE USE AMONG MELBOURNE DRIVERS: A
PREVENTABLE EXPOSURE TO INJURY RISK
Author(s): TAYLOR DM / BENNETT DM / CARTER M
Source: MEDICAL JOURNAL OF AUSTRALIA V.179 NO.3, PP.140-142,
4 AUG 2003

S 03 - 0842
WIRELESS TELEPHONES AND THE RISK OF ROAD CRASHES
Author(s): NADEAU CL / MAAG U / BELLAVANCE F
Source: ACCIDENT ANALYSIS & PREVENTION V.35 NO.5, PP.649-
660, SEP 2003

S 03 - 0789
THE DISTRACTION EFFECTS OF PHONE USE DURING A CRUCIAL
DRIVING MANEUVER
Author(s): HANCOCK PA / LESCH M / SIMMONS L
Source: ACCIDENT ANALYSIS & PREVENTION V.35 NO.4, PP.501-
514, JUL 2003

S 03 - 0788
EFFECT OF CELLULAR TELEPHONE CONVERSATIONS AND
OTHER POTENTIAL INTERFERENCE ON REACTION TIME IN A
BRAKING RESPONSE
Author(s): CONSIGLIO W / DRISCOLL P / WITTE M
Source: ACCIDENT ANALYSIS & PREVENTION V.35 NO.4, PP.495-
500, JUL 2003

S 03 - 0784
THE EFFECT OF CELL PHONE TYPE ON DRIVERS SUBJECTIVE
WORKLOAD DURING CONCURRENT DRIVING AND
CONVERSING
Author(s): MATTHEWS R / LEGG S / CHARLTON S
Source: ACCIDENT ANALYSIS & PREVENTION V.35 NO.4, PP.451-
457, JUL 2003

S 03 - 0782
EFFECTS OF TAIWAN IN-VEHICLE CELLULAR AUDIO PHONE
SYSTEM ON DRIVING PERFORMANCE
Author(s): LIU YC
Source: SAFETY SCIENCE V.41 ISSUE 6, PP.531-542, JUL 2003

S 03 - 0589
PILOT STUDY OF DISTRACTED DRIVERS
Author(s): GLAZE AL / ELLIS JM
Source: VIRGINIA COMMONWEALTH UNIV., RICHMOND. CENTER
FOR PUBLIC POLICY, 63PP. JAN 2003
URL: http://www.vcu.edu/uns/Releases/2003/march/DistractedReport.pdf

S 03 - 0462
COLLISION AND VIOLATION INVOLVEMENT OF DRIVERS WHO
USE CELLULAR TELEPHONES
Author(s): WILSON J / FANG M / WIGGINS S
Source: TRAFFIC INJURY PREVENTION V.4 ISSUE 1, PP.45-52, MAR
2003

S 03 - 0385
THE IMPACT OF HANDS-FREE MESSAGE RECEPTION/RESPONSE
ON DRIVING TASK PERFORMANCE
Author(s): COOPER PJ / ZHENG Y / RICHARD C
Source: ACCIDENT ANALYSIS & PREVENTION V.35 NO.1, PP.23-35,
JAN 2003

S 03 - 0362
CELL PHONE USE CAN LEAD TO INATTENTION BLINDNESS
BEHIND THE WHEEL
Author(s): STRAYER DL / DREWS FA / JOHNSTON WA

Source: NSC INJURY INSIGHTS NEWSLETTER, PP.1,4,7, FEB/MAR
2003

S 03 - 0301
CELL PHONE-INDUCED FAILURES OF VISUAL ATTENTION
DURING SIMULATED DRIVING
Author(s): STRAYER DL / DREWS FA / JOHNSTON WA
Source: JOURNAL OF EXPERIMENTAL PSYCHOLOGY: APPLIED V.9
NO.1, PP.23-32, 2003

S 03 - 0470
CELL PHONE USE WHILE DRIVING IN NORTH CAROLINA: 2002
UPDATE REPORT
Author(s): STUTTS JC / HUANG HF / HUNTER WW
Source: NORTH CAROLINA UNIV., CHAPEL HILL. HIGHWAY
SAFETY RESEARCH CENTER, 58PP. DEC 2002
URL: http://www.hsrc.unc.edu/pdf/2001/cellphone.pdf

S 03 - 0074
COGNITIVE DEMANDS OF HANDS-FREE PHONE CONVERSATION
WHILE DRIVING
Author(s): NUNES L / RECARTE MA
Source: TRANSPORTATION RESEARCH PART F: TRAFFIC
PSYCHOLOGY AND BEHAVIOUR V.5 NO.2, PP.133-144, JUN
2002

S 02 - 1508
DRIVERS' USE OF HAND-HELD CELL PHONES BEFORE AND
AFTER NEW YORK STATE'S CELL PHONE LAW
Author(s): MCCARTT AT / BRAVER ER / GEARY LL
Source: INSURANCE INST. FOR HIGHWAY SAFETY, ARLINGTON,
VA, 14PP. AUG 2002

S 02 - 1474
DISTRACTIONS AND THE RISK OF CAR CRASH INJURY: THE
EFFECT OF DRIVERS' AGE
Author(s): LAM LT
Source: JOURNAL OF SAFETY RESEARCH V.33 NO.3, PP.411-419,
FALL 2002

S 02 - 1337
THE RISK OF USING A MOBILE PHONE WHILE DRIVING
Author(s): NO*AUTHOR*
Source: ROYAL SOCIETY FOR THE PREVENTION OF ACCIDENTS,
    BIRMINGHAM, ENGLAND, 35PP. 2002
URL: http://www.rospa.org.uk/pdfs/road/mobiles/report.pdf

S 02 - 1087
THE DISCONNECT BETWEEN LAW AND POLICY ANALYSIS: A
    CASE STUDY OF DRIVERS AND CELL PHONES
Author(s): HAHN RW / DUDLEY PM
Source: AEI-BROOKINGS JOINT CENTER FOR REGULATORY
    STUDIES, WASHINGTON, DC, 56PP. MAY 2002
URL: http://aei.brookings.org/admin/pdffiles/working_02_07.pdf

S 02 - 0834
EXTENT AND EFFECTS OF HANDHELD CELLULAR TELEPHONE
    USE WHILE DRIVING (RESEARCH REPT. SEP 99-OCT 00)
Author(s): CRAWFORD JA
Source: TEXAS TRANSPORTATION INST., COLLEGE STATION,
    254PP. FEB 2001
URL: http://swutc.tamu.edu/Reports/167706-1.pdf

S 02 - 0522
CELL PHONE USE WHILE DRIVING IN NORTH CAROLINA
Author(s): REINFURT DW / HUANG HF / FEAGANES JR
Source: NORTH CAROLINA UNIV., CHAPEL HILL. HIGHWAY
    SAFETY RESEARCH CENTER, 31PP. NOV 2001
URL: http://www.hsrc.unc.edu/pdf/2001/cellphone.pdf

S 02 - 0050
THE IMPACT OF AUDITORY TASKS (AS IN HANDS-FREE CELL
    PHONE USE) ON DRIVING PERFORMANCE
Author(s): NO*AUTHOR*
Source: ICBC TRANSPORTATION SAFETY RESEARCH, BRITISH
    COLUMBIA, CANADA, 12PP. JUN 2001
URL: http://www.icbc.com/Library/research_papers/Cell_phones/
    Cellphones_Impact2.pdf

S 02 - 0040
DRIVEN TO DISTRACTION: DUAL-TASK STUDIES OF SIMULATED
DRIVING AND CONVERSING ON A CELLULAR PHONE
Author(s): STRAYER DL / JOHNSTON WA
Source: PSYCHOLOGICAL SCIENCE V.12 NO.6, 5PP. NOV 2001
URL: http://www.psych.utah.edu/AppliedCognitionLab/PS-Reprint.pdf

S 01 - 0926
ISSUES IN THE EVALUATION OF DRIVER DISTRACTION
ASSOCIATED WITH IN-VEHICLE INFORMATION AND
TELECOMMUNICATIONS SYSTEMS
Author(s): TIJERINA L
Source: TRANSPORTATION RESEARCH CENTER, EAST LIBERTY,
OH, 13PP. 2001
URL: http://www-nrd.nhtsa.dot.gov/departments/nrd-13/driver-distraction/
PDF/3.PDF

S 01 - 0941
CAN COLLISION WARNING SYSTEMS MITIGATE DISTRACTION
DUE TO IN-VEHICLE DEVICES?
Author(s): LEE JD / RIES ML / MCGEHEE DV
Source: NATL. HIGHWAY TRAFFIC SAFETY ADMIN.,
WASHINGTON, DC, 9PP. 2000
URL: http://www-nrd.nhtsa.dot.gov/departments/nrd-13/driver-distraction/
PDF/31.PDF

S 01 - 0131
THE INFLUENCE OF THE USE OF MOBILE PHONES ON DRIVER
SITUATION AWARENESS
Author(s): PARKS A / HOOIJMEIJER V
Source: TRANSPORTATION RESEARCH LAB., CROWTHORNE,
ENGLAND, 9PP. 2000
URL: http://www-nrd.nhtsa.dot.gov/departments/nrd-13/driver-distraction/
PDF/2.PDF

S 00 - 0552
CELLULAR PHONE USE WHILE DRIVING: RISKS AND BENEFITS
Author(s): LISSY KS / COHEN JT / PARK MY
Source: HARVARD CENTER FOR RISK ANALYSIS, BOSTON, MA,
73PP. JUL 2000

S 02 - 0049
INVESTIGATION OF THE USE OF MOBILE PHONES WHILE
DRIVING
Author(s): CAIN A / BURRIS M
Source: CENTER FOR URBAN TRANSPORTATION RESEARCH,
TAMPA, FL, 46PP. APR 1999
URL: http://www.cutr.eng.usf.edu/its/mobile_phone.htm

S 01 - 0839
THE ECONOMICS OF REGULATING CELLULAR PHONES IN
VEHICLES
Author(s): HAHN RW / TETLOCK PC
Source: AEI-BROOKINGS JOINT CENTER FOR REGULATORY
STUDIES, WASHINGTON, DC, 40PP. OCT 1999
URL: http://www.aei.brookings.org/publications/working/
working_99_09.pdf

S 98 - 0713
CELLULAR PHONES AND FATAL TRAFFIC COLLISIONS
Author(s): VIOLANTI JM
Source: ACCIDENT ANALYSIS & PREVENTION V.30 NO.4, PP.519-
524, JUL 1998

S 98 - 0545
INVESTIGATION OF THE SAFETY IMPLICATIONS OF WIRELESS
COMMUNICATIONS IN VEHICLES
Author(s): NO*AUTHOR*
Source: NATL. HIGHWAY TRAFFIC SAFETY ADMIN.,
WASHINGTON, DC, 278PP. 1997
URL: http://www.nhtsa.dot.gov/people/injury/research/wireless/

S 97 - 0117
ASSOCIATION BETWEEN CELLULAR-TELEPHONE CALLS AND
MOTOR VEHICLE COLLISIONS
Author(s): REDELMEIER DA / TIBSHIRANI RJ
Source: NEW ENGLAND JOURNAL OF MEDICINE V.336 NO.7,
PP.453-458, 13 FEB 1997

INDEX

DATE DUE

MAY 1 4 2013			